Consciousness as a Scientific Concept

Studies in Brain and Mind

VOLUME 5

For further volumes:
http://www.springer.com/series/6540

Elizabeth Irvine

Consciousness as a Scientific Concept

A Philosophy of Science Perspective

Springer

Elizabeth Irvine
Centre for Integrative Neuroscience
University of Tuebingen
Germany

ISBN 978-94-007-5172-9 ISBN 978-94-007-5173-6 (eBook)
DOI 10.1007/978-94-007-5173-6
Springer Dordrecht Heidelberg London New York

Library of Congress Control Number: 2012944703

Printed on acid-free paper

Springer is part of Springer Science+Business Media (www.springer.com)

Preface

In this book, I argue for an eliminativist stance towards the concept of consciousness as it is used in scientific research. Given the recent rise of consciousness science, and the associated rise in philosophical engagement with this field of research, this is clearly a controversial stance to try to defend. However, while some colleagues have viewed my claims as fundamentally misguided, others have viewed them as entirely sensible and rather unsurprising. A lot seems to depend on the background assumptions that are brought to the table. Therefore, while I do not count on converting many from the first camp, I hope that the material presented here at least points to methodological problems in consciousness science that deserve more attention. I also hope that it spurs more use of philosophy of science in consciousness studies, as I think it mas much to offer current debates.

The approaches taken in different parts of this book stem from different academic influences, and while I do not claim to have convinced those who have inspired the methods used here, I am grateful to them for their support and discussions. Much of the first part of the book about measures of consciousness comes from discussions with Michael Snodgrass, who introduced me to the wonderful world of Signal Detection Theory. The mechanistic approach developed from time spent at the University of Pittsburgh as a graduate student. Discussions about property clusters, scientific kinds, and related eliminativist claims with the late Barbara Scholz influenced many different sections of the book. I have also had many enlightening discussions about scientific eliminativism with Aaron Sloman. The framework of treating perception as a generative process comes from discussions with Andy Clark. Ned Block has been kind enough to discuss all my criticisms of his work, and has been very supportive despite our near total disagreement about the outlook of consciousness science. Andy Clark, Rob McIntosh and John Henderson oversaw much of what is written here, and provided essential guidance in making it all into a coherent whole.

More generally, my thanks go to: the PPLS research community at the University of Edinburgh, particularly the PPIG group and work in progress audiences for stimulating interdisciplinary discussions; my new research community at the CIN in Tuebingen for help with Chap. 6; and special thanks go to my colleague Matteo Colombo for all his incisive comments on many of the chapters here. Sean Roberts has provided ongoing encouragement, support and bad puns over the last few years, as well as editorial and technical assistance in preparation of the book, for which I am very grateful.

Finally, thanks the British Society for Philosophy of Science for funding my PhD, and making this research possible. Also, many thanks to Mark Sprevak, Gualtiero Piccinini, Edouard Machery, an anonymous reviewer of the book manuscript, and Ties Nijssen at Springer, for all their support in preparing this book.

Notes on Permissions

Contents

List of Abbreviations

2AFC	2 alternative forced-choice
ERP	Event related potential
GNWT	Global Neuronal Workspace theory
HPC	Homeostatic property cluster
INK	Investigative Kind concept
LTP	Long-term potentiation
NCCs	Neural correlates of consciousness
RP	Recurrent processing
SDT	Signal Detection Theory

Chapter 1
The Scientific Study of Consciousness

1.1 (Anti-) Introduction

Consciousness is currently a hot topic in both philosophy and in science, and it is a notoriously difficult one. We are all supposed to be intimately familiar with the phenomenon of consciousness, yet there are a surprisingly wide range of views about how to characterise it, and how to investigate it. In the last 20 years or so, researchers from a range of scientific fields have attempted to create a science of consciousness. Assessing the viability of this science of consciousness is the subject of this book, pursued from the viewpoint of philosophy of science. This is a novel approach, and one that is best introduced by contrasting it with more traditional philosophical approaches towards consciousness science. These approaches are outlined very briefly below, followed by an introduction to the methods used throughout the rest of the book.

Philosophers tend to be divided about just what a science of consciousness can achieve. Philosophical arguments against the possibility of there being a complete scientific theory of consciousness are typically based on Levine's (1983) 'Explanatory Gap'. This refers to the gap between knowledge of the physical world and knowledge about the phenomenal world, or the world of experience. Arguments based on the explanatory gap state that whatever scientific theory of consciousness we get, it will leave out something essential: the 'felt' qualities or the 'what-it-is-like-ness' of experience.

Using this intuition, Chalmers (1995) has identified two types of problems related to consciousness, one of which he argues that science can answer, and the other that science cannot. Chalmers acknowledges that 'consciousness' refers to many different cognitive, neurophysiological and sensory phenomena, and that each of these can be investigated scientifically. He claims that questions about these phenomena form the 'easy problems' of consciousness, including questions about the neural basis of reportability, the neurophysiological differences between sleep and wakefulness, how sensory systems work, how complex cognitive processing is achieved, and so on. While not particularly easy in scientific terms, these problems are clearly scientifically tractable ones.

E. Irvine, *Consciousness as a Scientific Concept,* Studies in Brain and Mind 5,
DOI 10.1007/978-94-007-5173-6_1,

In contrast, the 'hard problem' is the one that the explanatory gap exposes. This is the problem of how and why experiences come from arrangements of physical entities. So, even if we know all there is to know about reportability, attention and visual processing, this won't tell us what its like to see a vibrant, busy, visual scene, or why such an experience arises from physical matter.

Block (1990, 1992, 1995) has also used this intuition to distinguish between those aspects of consciousness that we can investigate through reports and behaviours—'access consciousness', and the subjective aspect of consciousness that science cannot be used (at least in a direct way) to investigate—'phenomenal consciousness'. That is, although we can investigate how people tend to react to colours and how colour vision works, our knowledge of these states does not allow us to infer how these outward behaviours relate to internal colour experiences. More recently Block has described ways in which scientific methodology can be used to investigate phenomenal consciousness (e.g. Block 2005, 2007), but again these are not directly based on investigations of reportability or other cognitive capacities.

There are of course critics of the distinction between the easy and hard problems of consciousness, and the distinction between access and phenomenal consciousness. Dennett (1991, 1996) argues that the hard problem arises out of conceptual confusion around the 'what-it-is-like-ness' of experience. He claims that there is nothing more to consciousness than just all those varied 'easy problems', and that Chalmers' argument that they leave something out (the 'what-it-is-like-ness') is similar to claiming that modern biology still can't explain why things are alive. He claims that the hard problem is simply incoherent, so a science of the easy problems of consciousness will provide a *complete* theory of consciousness.

Scientific researchers themselves vary in how they treat the distinction between the easy and hard problems of consciousness. Some accept the idea that all they can really do is to investigate access consciousness, or the easy problems, and that whatever they do leaves the hard problem and phenomenal consciousness untouched. Faced with arguments from Chalmers or Block, they argue that this is all science is equipped to do, so it is all that they are concerned with: "Given the lack of scientific criterion, at this stage at least, for defining conscious processing without reportability, the dissociation between access and phenomenal consciousness remains largely speculative and even possibly immune to scientific investigation" (Kouider et al. 2007, p. 2028). The strength of scientific claims, so the strength of scientific language, is often affected by philosophical arguments. For example, the project to identify the neural states that co-vary with the contents of behaviours or reports is called the Neural Correlates of Consciousness project. This term is used to underline the idea that mental states cannot be *identified* with physical states, only correlated, a much weaker claim. Alternatively, some researchers go head on to try to investigate phenomenal consciousness, for example as recurrent processing (Lamme 2006), or by mapping out qualia-spaces in terms of informational relationships (Tononi 2008).

This very brief outline of some of the central philosophical thinking about the possibility and limits of a science of consciousness, and how this is interpreted by scientists, illustrates one way of tackling questions about a science of consciousness.

That is, the work of Chalmers, Block and Dennett shows how the science of consciousness can be evaluated on philosophical or conceptual grounds. However, this philosophical literature is *not* presented in order to introduce the method or contents of this book. Instead, it is presented above in order to contrast it with the very different methods used throughout the following chapters. Although questions will still be asked about the validity of distinctions like access and phenomenal consciousness, and the relation between cognitive abilities and functional roles, it will be done in a different way to that usually encountered.

Rather than looking from the outside in, and making claims about the possible questions that science can or cannot answer, the claims made in this book are based on an investigation of the methods and results of contemporary consciousness science. In particular, it will be questioned whether consciousness science is a 'good' science in terms of its theoretical and experimental practices, and thus whether the concept of consciousness itself is a scientifically viable concept.[1] This will be done by evaluating whether specific research methods are applied appropriately, and whether standard research heuristics guide research in appropriate ways. If they are not, this points to problems in the basic structure of consciousness science, and (as I will argue) with the status of 'consciousness' as a viable scientific concept. In order to begin this investigation, it is first necessary to look at the recent history, goals, methods and theories of contemporary consciousness science.

1.2 A Brief History of Consciousness Science

The history of research into consciousness is not a straightforward one (Dixon 1971). Pre-empting the current focus on visual consciousness, consciousness research from the late nineteenth century onwards was carried out largely through psychophysical experiments of sensory perception. Along with James' (1890) general support for the use of introspection as a means to assess awareness, early researchers used subjects' reports as a measure of awareness.

For example, Sidis (1898) determined the distance from a stimulus at which subjects claimed to see only a faint spot, and then tested their performance on an alphanumeric discrimination task. Although subjects reported seeing only faint spots, they still performed better than chance at the discrimination task. Confidence ratings were also used to assess awareness. Pierce and Jastrow (1884) tested subjects' ability to tell the difference between small weight increases or decreases on a weight on their finger, using judgements made along a four-point confidence rating scale, to assess the presence or absence of consciousness. Under the assumption that a low confidence rating meant that subjects were not conscious of a particular weight change, they also found that subjects were able to accurately discriminate weight increases from decreases in the absence of consciousness. Experiments like these

[1] 'Consciousness' will be used throughout to denote the concept of consciousness.

resulted in claims about the existence of a wide range of unconscious perceptual abilities.

However, given the failure of introspective methods to provide psychology with laws and theories about consciousness and mental life more generally, and its inherent methodological problems (e.g. Dunlap 1912), subjective approaches were increasingly rejected as a viable method in psychology. Discussed in more detail below, the application of Signal Detection Theory to human perceptual systems from the early 1950s onwards (see e.g. Blackwell 1952; Eriksen 1960; Goldiamond 1958; Green and Swets 1966) also showed that reports are highly manipulable and context-sensitive, so were argued to be unreliable ways of assessing awareness or perceptual discrimination.

Signal Detection Theory showed that reports are based both on a subjects' underlying ability to discriminate stimuli (sensitivity), and the 'response criterion' of the subject. The response criterion is a threshold set according to task and context that determines the strength of perceptual information required to make a particular response by a subject. It is therefore possible that subjects can perceive stimuli even if they do not 'decide' to report them. Thus, it was argued that subjective measures based on reports were not simple measures of awareness (or perceptual abilities), but indications of how subjects set their response criteria. To overcome the problems in using reports, which may underestimate the stimulus features that subjects are conscious of, it was proposed that a measure of the underlying ability to discriminate stimuli, an objective measure, should be used instead.

Yet even with these new measures, the rise of behaviourism, with its focus away from mental states (e.g. Skinner 1953; Watson 1913), meant that consciousness was rejected as a suitable phenomenon for scientific research for many years. Even with the rise of cognitivism in the 1960s (e.g. Chomsky 1959), consciousness was still not a topic that many researchers thought an appropriate one for the mind sciences. This was because of the lack of a clear computational structure or functional role that consciousness could be equated with. However, consciousness was often implicitly assumed to be identical with attention and investigated under this research program, a continuing but controversial trend in contemporary consciousness science (Baars 1988, 1997; Block 2005; Dehaene et al. 2006; Koch and Crick 2004; Lamme 2004; Mack and Rock 2003; Prinz 2005).

From the 1990s onwards, partly due to new experimental technology and techniques for investigating cognitive abilities and their underlying mechanisms, consciousness research was again seen as a viable field. These technologies made it possible to investigate brain function in a non-invasive way, without having to rely on existing pathologies or brain lesions in human subjects. The ability to investigate the neural mechanisms in the brain made it seem possible that researchers could investigate the relationship between physical processes and consciousness.

How researchers now view the problem of consciousness is summarised in Crick and Koch's (1990) seminal paper that set out the current research agenda. Here, they outline two basic assumptions; that there is something that can be scientifically explained, and the different ways that consciousness is manifest are due to one (or a few) common mechanisms. They also identify the main research questions for a science of consciousness, including the binding problem, what sort of mechanism the mechanism for consciousness is, the methodological problem of separating a mechanism for consciousness from necessary background conditions or other cognitive processes, and the role of attention and short term memory in determining the contents of consciousness.

As a result of this new research program, many disparate fields of research can now be seen as part of consciousness science. Research into the electrophysiology of sleep and wake cycles in the medical domain, (Alkire et al. 2008; Gottesmann 1999; Nir and Tononi 2010), research into implicit and explicit learning (Cleeremans 2008), priming (Kouider et al. 2007) and inhibition (Jacoby 1991; Visser and Merikle 1999), and the functioning of the visual system, including its relation to motor areas (Milner and Goodale 2008), all now contribute to consciousness science. All of this varied research has culminated in roughly three types of scientific theories of consciousness, described below.

1.3 Current State of the Science: Theories and Taxonomies of Consciousness

There are currently a wide range of theories of consciousness and associated taxonomies of conscious states. In their (2008) review, Seth et al. divide these theories into worldly discrimination theories, integration, and higher order thought theories of consciousness. The general claims made within these theories and the problems they face are noted below, followed by a brief description of some of the common distinctions made between different types of consciousness.

First, worldly discrimination theories of consciousness state that consciousness is exhibited through behaviours that show a subjects' ability to detect or discriminate stimuli. These theories are typically based on objective measures that provide performance-based ways of assessing the presence or absence of consciousness, including the measure d' defined within Signal Detection Theory (abbreviated to SDT). As SDT has far reaching implications in consciousness research it is described in more detail here.

As noted above, reports were originally used as markers of consciousness. However, the application of Signal Detection Theory to human perceptual systems showed that reports are subject to response biases, and may not therefore accurately reflect the amount of information available and reportable for a subject. An example of the application and implications of Signal Detection Theory (SDT) is given in terms of the phenomenon of perceptual defence (for original papers on perceptual defense see e.g. Bruner and Postman 1947a, b, 1949). Here, two sets of words are flashed at subjects, one set is neutral (e.g. 'shot') and one might be swear words (e.g. 'shit') or sexually loaded words. Despite both sets of words being shown under equivalent conditions, subjects are consistently better at freely reporting or responding to the neutral words than the swear words. Researchers claimed that the swear words were perceived unconsciously, and then repressed by some defence mechanism.

However, early SDT theorists argued that both sets of words are processed to an equally high level, but that subjects do not like reporting swear words. That is, they have a higher internal threshold or criterion level for reporting 'shit' than for 'shot'. This is because subjects will probably be more worried about false positives (reporting the presence of a stimulus when it is not there) for swear words than for neutral words. Subjects want to be very confident that they see 'shit' before they report it, but will report lower confidence perceptions of 'shot'. Perceptual defense illustrates how, due to response bias, subjective reports can be unreliable measures of what information a subject has available and is capable of reporting.

From this, objective measures that are free from response bias became more popular measures of conscious perception. Derived from SDT, the measure d' is a measure of a subjects' ability to discriminate signals (target stimuli) from noise (external sensory noise or noisy internal processing). Accordingly, the measure d' is referred to as a measure of a subjects' *sensitivity* to stimuli. The objective (detection) threshold $d' = 0$ is defined as the threshold at which subjects can no longer detect signals from noise above chance level in forced-choice tasks (i.e. when subjects are forced to choose between alternative options). This threshold differs across tasks, but is reasonably similar across subjects and stable over time. In contrast, the threshold at which subjects freely report detection of signals, the subjective (detection) threshold, is also determined by the response criterion β of the subject, which can vary wildly according to variables such as task type, type and length of training, and motivation (Green and Swets 1966).

According to worldly discrimination theories of consciousness, if subjects fail to report a stimulus in conditions above $d' = 0$, this report is seen as a product of response bias, not a sign that subjects are necessarily unconscious of the stimuli. Another crucial part of SDT is that perception is graded both by the performance rates for particular tasks, but also by different types of tasks. Aside from detection, there are other levels of information that are used by subjects in identification, categorisation, discrimination, recognition tasks, as well as different types of confidence ratings.

Based on the application of SDT to subjects' responses, the objective measure d' has been used to index conscious from unconscious perception, and is still often used in investigations of subliminal, or unconscious perception (Kouider and Dehaene 2007). However, whether the subjective or objective measures are better measures of consciousness is still a highly debated topic in consciousness studies (see Snodgrass et al. 2004 and replies). This debate, as well as the gradations in perception identified within SDT, will feature heavily in this book, particularly in the first few chapters that follow this introduction.

In contrast to worldly discrimination theories, integration theories are based on the idea that consciousness depends on the ability to integrate and share information across brain areas. According to integration theories, consciousness plays an executive, selective, controlling role that is made possible by the sharing of information from sensory areas to areas involved in planning, decision-making, and action. Baars' Global Workspace model (Baars 1988, 1997; Shanahan and Baars 2007), Dehaene's Neuronal Workspace model (Dehaene and Changeux 2004; Dehaene and Naccache 2001; Dehaene et al. 2006), and Tononi's computational Information

Integration Theory (2004, 2008) are all based on the equation between consciousness and the global availability of information. Across these theories, information is made available through attentional selection, recurrent processing, neural synchrony, or can be characterised in computational terms. The scope of integration theories across psychological, neurophysiological and computational models of consciousness illustrates how widespread this conception of consciousness is. However, while integration theories suggest that conscious behaviours may be more complex than those found in worldly discrimination theories, experimental work on integration theories is sometimes based on the same objective measures of behaviour (e.g. continued use of d' in investigating unconscious perception). Therefore, while integration theories of consciousness appear to better capture the complexity of conscious states, they sometimes make use of the same behavioural measures used in worldly discrimination theories.

Higher order thought theories of consciousness (HOT) are different again. They are based on the idea that if someone is conscious of something, this means that they are aware of a representation of it. This idea, originally introduced by Rosenthal (e.g. 1993, 2005), has been altered slightly in the scientific literature, so HOT theories are now taken to refer to theories in which being conscious of something entails that subjects are able to comment on this state (e.g. by producing confidence judgements about seeing it), or that being conscious of something entails having some disposition or attitude towards it. For example, Cleeremans' Radical Plasticity Thesis (2008) suggests that consciousness of x consists of all the emotional, remembered, and behavioural associations we have with x, which together give us an 'attitude' towards x and constitutes our consciousness of it. Lau (2008) suggests that consciousness of x is reflected in a subject's ability to generate appropriate commentaries, in the form of second order confidence ratings, towards their ability to discriminate x. That is, subjects are conscious of something if they are able to correctly judge how accurate their responses are towards it (similar to Pierce and Jastrow's method discussed above). Although HOT theories describe consciousness in a different way to worldly discrimination theories and integration theories, they are again based on questionable assumptions about the adequacy of the measures they use (discussed in Chap. 2).

The similarities and differences between different theories and measures of consciousness are reflected in the main taxonomies of conscious states. By far the most cited is Block's (1990, 2001, 2007) distinction between phenomenal and access consciousness (and sometimes reflective consciousness). Phenomenal consciousness, as originally defined, is the subjective aspect of consciousness, and argued to be non-functionalisable. Access consciousness is the aspect that can be captured in functional terms. Phenomenal consciousness is now used to refer to *states* of consciousness that cannot be probed using behavioural measures, and access consciousness to those states that can be probed. Reflective consciousness refers to a subset of access consciousness that involves metacognition or higher order awareness of first order states (such as those found in HOT theories). Although behavioural measures underlie all theories of consciousness, Block (see e.g. 2005, 2007) and Lamme (2004, 2006) argue that there is room for phenomenal consciousness in non-reported, non-integrated, and non-reflective processing, which has generated much debate over whether such states of consciousness can exist or are scientifically investigable.

Another distinction often referred to in consciousness studies is the difference between creature and state consciousness (for application see e.g. Laureys 2005). This distinction captures two of the meanings of consciousness. Creature consciousness refers to the state of being conscious at all. State consciousness refers to cases when a subject is conscious *of* something. The Neural Correlates of Consciousness (NCC) project tends to focus on establishing the neural correlates for state consciousness, such as the neural correlate for conscious perception of colour or motion. Accordingly, most of the content of this book will focus on research into state consciousness, though in practice the distinction between state and creature consciousness is far from clear. For example, integration theories in particular make claims about the general conditions for consciousness at the same time as suggesting how specific cases of consciousness arise. There is also much debate over how to demarcate NCCs such that background conditions of creature consciousness are left out while central components of particular instances of state consciousness are included (e.g. Chalmers 2000).

1.4 Assessing the Science of Consciousness

Having provided a brief description of what the book is *not* about, and about the history and current state of consciousness research, it is now possible to describe the motivating question of the following chapters and how it will be approached. The central question is whether consciousness science is a viable science, according to the standard practices found elsewhere in science, particularly as formulated in philosophy of psychology, biology and neuroscience. These practices include applications of dissociation logic, ways of integrating research to provide convergent evidence for theories and measures, ways of identifying neural mechanisms, and the way that cross-level correlation (identity) claims are put forward between the contents of consciousness and contents of sensory or cognitive processing. These methods are all common ones in science, but come with a range of conditions that must be met if they are to be properly applied and provide any results of scientific merit. By looking in depth at the way these methods are used in consciousness science, in terms of experimental design, the interpretation of data, and common assumptions used, I aim to establish whether these methods are used appropriately. Where they are not, I aim to show how they could be used appropriately, and what conclusions to draw given their proper application.

Scientific methods also typically fulfill a heuristic role. In this role, methods are used not only to answer current research questions, but also to provide guidance in how to formulate new research questions, and to assess the utility of current conceptual schemes. For example, dissociation logic as used in psychology provides support for hypotheses about the structure of cognitive systems, but the products of dissociation paradigms are also used in an iterative pattern to refine hypotheses, suggest new experimental paradigms, and revise the concepts used to interpret experimental findings.

Likewise, assessing the truth of cross-level identity claims is useful in its own right, but the ways in which identity claims fail also suggest ways of revising concepts at both levels of description, through which a better supported identity claim can be made. As well as investigating whether scientific methods are appropriately applied in consciousness science, I also investigate whether they fulfill their standard (and crucial) heuristic roles. Again, if they do not, I will suggest what sorts of conceptual clarifications are necessary if their heuristic role is taken seriously.

Given these investigations, questions can then be raised over the validity of 'consciousness' as a scientific concept. For example, it can be questioned if 'consciousness' refers to any phenomena about which reliable predictions and broad generalisations can be made (i.e. if 'consciousness' refers to any scientific kinds). If not, it can be questioned if concepts of consciousness can be used in guiding and stating research goals. However, even if the concept of 'consciousness' cannot be used in either of these ways, there may yet be pragmatic reasons for continuing to use concepts of consciousness. These reasons, including whether concepts of consciousness can be used unambiguously, and whether they are essential to scientific communication and research continuity, can be assessed to see if they make a science of consciousness acceptable on pragmatic, if not ideal, grounds.

Looking at the application of different scientific methods and their heuristic roles in consciousness science makes it possible to question whether the current science of consciousness is satisfactory in its own right. This reverses the usual strategy of investigating the limits of a science of consciousness from a philosophical or conceptual viewpoint. Typically, philosophers assume some conception of consciousness and then argue whether or not (or to what extent) science can be successful in generating a theory of consciousness. The approach used here is a more directly naturalistic one. That is, the strategy is to see how the presence of concepts of consciousness in a science affects the ways that methods and research heuristics are applied. If the presence of concepts of consciousness mean that standard scientific methodologies are invalidly applied, and their heuristic role ignored, then these concepts should not figure in science. Instead, scientific research into our sensory and cognitive capacities can be used to pinpoint where intuitions and confusions about concepts of consciousness stem from, and can themselves be explained.

The position reached at the end of this investigation is an eliminativist one, but the route to this position is a new one. Other eliminativist accounts use conceptual analysis, analogies from the history of science, as well as empirical work in psychology and neuroscience, but none of them stem from an in-depth consideration of the actual practices of contemporary consciousness science. Thus, Wilkes (1984, 1988) argues that 'consciousness' does not refer to any natural kinds based on linguistic considerations. Sloman 2010 (Sloman and Chrisley 2003) argues that an explanation and theory of 'consciousness' consists of just those explanations and theories of the abilities we typically associate with 'consciousness', such as self-monitoring, affective responses, and so on. Dennett argues that 'consciousness' can be explained by reference to the emergent behaviours of a set of dumb, competing parallel processes (his Multiple Drafts account, Dennett 1991). In contrast, the aim here is to show, by taking the intuitions that lie behind the science of consciousness seriously, that concepts of consciousness present serious problems in the practice of science. It will be argued that these concepts are so deeply flawed that not even the easy problems of consciousness make sense as problems *of consciousness*. Instead, they are fairly hard problems about a wide array of cognitive capacities and processes. The structure of the following chapters is given below.

1.5 The Plan

The first section of the book examines the methodological difficulties in establishing whether subjective measures (such as subjects' reports) or objective measures (such as task performance) are better measures of consciousness. Much of this debate can be traced to the history of psychology and its preferred methods as mentioned above, but the full extent of the methodological problems are explained in this first section. Chapter 2 includes an investigation of two subjective measures of consciousness based on introspective techniques. They are analysed in light of the arguments against the use of subjective, introspective methods that were first raised up to a century ago, as well as responses from SDT. A different type of subjective measure is then assessed that is supposed to overcome these problems, but fails for different reasons. This suggests that subjective measures of consciousness do not currently offer an acceptable way of investigating consciousness.

The third and fourth chapters contain discussions of objective measures (based on task performance), and the basic methodological problems in establishing *any* measure of consciousness. Consciousness researchers look for qualitative differences in behaviour and use dissociation techniques to identify consciousness, but it will be questioned whether these methods are appropriately applied in consciousness research. It is also questioned whether the heuristic role of these methods is preserved in consciousness research, and if not, then what would be the likely result if it were fulfilled.

Chapter 5 questions the use of another popular method in science, through which the integration of various approaches is used to provide convergent evidence for a measure or hypothesis. This method has been proposed by Seth et al. (2008) as a way of overcoming differences between the results of different behavioural and neurophysiological measures of consciousness. By combining multiple measures in single experimental paradigms, they suggest that it should be possible to establish how different measures relate to each other, and identify what exactly it is that they measure. The utility of such an approach in consciousness science is addressed by looking in detail at the debates between proponents of different neurophysiological measures of consciousness and the behavioural operationalisations that they are based on. By looking at the necessary pre-conditions on experimental practice for successful integration, it is possible to see whether integrative approaches hold any promise in consciousness science.

Chapter 6 addresses the use of another method found in neuroscientific research into consciousness. Neuroscience and neurophysiological research are used to try to identify the neural mechanisms of consciousness. The ways in which mechanisms are demarcated, and how the target of the mechanism is identified, are both subject to standards of scientific practice. The criteria for demarcating mechanisms that are appealed to in this chapter are described by Craver (2007). By exploring debates about the mechanism of consciousness it is possible to question if the demarcation of mechanisms of consciousness conform to these criteria, and if not, what their proper application would entail. As mechanisms have been used to describe scientific kinds (Boyd 1991; Kornblith 1993), the mechanisms that result from the proper application of demarcation criteria have a clear bearing on whether consciousness picks out any scientific kinds, part of an eliminativist claim.

The two chapters following Chap. 6 examine a method that is particular to consciousness science. This is content-matching strategy, in which the contents of consciousness are mapped to the contents of neural states or processes (part of this forms the neural correlates of consciousness project). First, a case study of one attempt to map the contents of phenomenal consciousness to the contents of sensory memory is used to highlight the differences between the structure and properties of the visual system, and the way that the contents of visual consciousness are characterised. This case study is extended to a more general investigation of the scope of this method in consciousness science. It is also questioned whether the heuristic role that inter-level identity claims typically play in science (McCauley and Bechtel 2001) is satisfied within the content-matching strategy.

In the final chapter all of these investigations into the methodological problems in consciousness science are used to argue that 'consciousness' is *not* a viable scientific concept. Criteria from other contemporary eliminativist accounts, including from Griffiths' (1997) evaluation of 'emotion', Machery's (2009) work on 'concept' in psychology, and debates about the elimination of 'species' from biology (Brigandt 2003; Ereshefsky 1998), are used to frame this final chapter. These accounts are all based on assessing the utility of these concepts in terms of how well they function in the practice of scientific research. This includes an assessment of whether they identify scientific kinds, their epistemic role, and their pragmatic value. That is, if the concept of consciousness can only play a negative role in scientific research, then its continued use in scientific research is problematic. The chapter ends by detailing the positive contributions this kind of eliminativist stance offers. A final conclusion provides a short summary of the approach taken in the book.

References

Alkire, M. T., Hudetz, A. G., & Tononi, G. (2008). Consciousness and anesthesia. *Science, 322*, 876–880.

Baars, B. J. (1988). *A cognitive theory of consciousness*. Cambridge: Cambridge University Press.

Baars, B. J. (1997). In the theatre of consciousness: Global workspace theory, a rigorous scientific theory of consciousness. *Journal of Consciousness Studies, 4*, 292–309.

Blackwell, H. R. (1952). Studies of psychophysical methods for measuring visual thresholds. *Journal of the Optical Society of America, 42*, 606–616.

Block, N. (1990). Consciousness and accessibility. *Behavioral and Brain Sciences, 13*, 596–598.

Block, N. (1992). Begging the question again phenomenal consciousness. *Behavioral and Brain Sciences, 15*, 205–206.
Block, N. (1995). On a confusion about a function of consciousness. *Behavioral and Brain Sciences, 18*, 227–287.
Block, N. (2001). Paradox and cross purposes in recent work on consciousness. *Cognition, 79*, 197–219.
Block, N. (2005). Two neural correlates of consciousness. *Trends in Cognitive Science, 9*, 46–52.
Block, N. (2007). Consciousness, accessibility, and the mesh between psychology and neuroscience. *Behavioral and Brain Sciences, 30*, 481–548.
Boyd, R. (1991). Realism, anti-foundationalism and the enthusiasm for natural kinds. *Philosophical Studies, 61*, 127–148.
Brigandt, I. (2003). Species pluralism does not imply species eliminativism. *Philosophy of Science, 70*, 1305–1316.
Bruner, J. S., & Postman, L. (1947a). Tension and tension-release as organizing factors in perception. *Journal of Personality, 15*, 300–308.
Bruner, J. S., & Postman, L. (1947b). Emotional selectivity in perception and reaction. *Journal of Personality, 16*, 69–77.
Bruner, J. S., & Postman, L. (1949). Perception, cognition, and behavior. *Journal of Personality, 18*, 14–31.
Chalmers, D. J. (1995). Facing up to the problem of consciousness. *Journal of Consciousness Studies, 2*, 200–219.
Chalmers, D. J. (2000). What is a neural correlate of consciousness? In T. Metzinger (Ed.) *Neural correlates of consciousness*. Cambridge: MIT Press.
Chomsky, N. (1959). A review of Skinner's verbal behaviour. *Language, 35*, 26–58.
Cleeremans, A. (2008). Consciousness: The radical plasticity thesis. In R. Banerjee & B. K. Chakrabarti (Eds.), *Progress in brain science, 168*, 19–33.
Craver, C. F. (2007). *Explaining the brain: Mechanisms and the mosaic unity of neuroscience*. Oxford: Clarendon Press.
Crick, F., & Koch, C. (1990). Towards a neurobiological theory of consciousness. *Seminars in the Neurosciences, 2*, 263–275.
Dehaene, S., & Changeux, J. P. (2004). Neural mechanisms for access to consciousness. In M. Gazzaniga (Ed.), *The cognitive neurosciences* (3rd ed., pp. 1145–1157). New York: Norton.
Dehaene, S., & Naccache, L. (2001). Towards a cognitive neuroscience of consciousness: Basic evidence and a workspace theory. *Cognition, 79*, 1–37.
Dehaene, S., Changeux, J. P., Naccache, L., Sackur, J., & Sergent, C. (2006). Conscious, preconscious, and subliminal processing: A testable taxonomy. *Trends in Cognitive Sciences, 100*, 204–211.
Dennett, D. C. (1991). *Consciousness explained*. London: Penguin.
Dennett, D. C. (1996). Facing backwards on the problem of consciousness. *Journal of Consciousness Studies, 3*, 4–6.
Dixon, N. F. (1971). *Subliminal perception: The nature of a controversy*. New York: McGraw-Hill.
Dunlap, K. (1912). The case against instrospection. *Psychological Review, 19*, 404–413.
Ereshefsky, M. (1998). Species pluralism and anti-realism. *Philosophy of Science, 65*, 103–120.
Eriksen, C. W. (1960). Discrimination and learning without awareness: A methodological survey and evaluation. *Psychological Review, 67*, 279–300.
Goldiamond, I. (1958). Indicators of perception: 1. Subliminal perception, subception, unconscious perception: An analysis in terms of psychophysical indicator methodology. *Psychological Bulletin, 55*, 373–411.
Gottesmann, C. (1999). Neurophysiological support of consciousness during waking and sleep. *Progress in Neurobiology, 59*, 469–508.
Green, D. M., & Swets, J. A. (1966). *Signal detection theory and psychophysics*. New York: Wiley.
Griffiths, P. E. (1997). *What emotions really are: The problem of psychological categories*. Chicago: University of Chicago Press.

Jacoby, L. L. (1991). A process dissociation framework: Separating automatic from intentional uses of memory. *Journal of Memory and Language, 30*, 513–541.

James, W. (1890). *The principles of psychology*. New York: Dover.

Koch, C., & Crick, F. (2004). The neuronal basis of visual consciousness. In L. M. Chalupa & J. S. Werner (Eds.), *The visual neurosciences* (pp. 1682–1694). Cambridge: MIT Press.

Kornblith, H. (1993). *Inductive inference and its natural ground: An essay in naturalisatic epistemology*. Cambridge: MIT Press.

Kouider, S., & Dehaene, S. (2007). Levels of processing during non-conscious perception: A critical review of visual masking. *Philosophical transactions of the Royal Society B, 362*, 857–875.

Kouider, S., Dehaene, S., Jobert, A., & Le Bihan, D. (2007). Cerebral bases of subliminal and supraliminal priming during reading. *Cerebral Cortex, 17*, 2019–2029.

Lamme, V. A. F. (2004). Separate neural definitions of visual consciousness and visual attention; a case for phenomenal awareness. *Neural Networks, 17*, 861–872.

Lamme, V. A. F. (2006). Towards a true neural stance on consciousness. *Trends in Cognitive Sciences, 10*, 494–501.

Lau, H. C. (2008). A higher order Bayesian decision theory of consciousness. In R. Banerjee & B. K Chakrabati (Eds.), *Progress in brain research, 168*, 35–48.

Laureys, S. (2005). The neural correlates of (un)awareness: Lessons from the vegetative state. *Trends in Cognitive Sciences, 9*, 556–559.

Levine, J. (1983). Materialism and qualia: The explanatory gap. *Pacific Philosophical Quarterly, 64*, 354–361.

Machery, E. (2009). *Doing without concepts*. Oxford: Oxford University Press.

Mack, A., & Rock, I. (2003). Inattentional blindness: A review. *Directions in Psychological Science, 12*, 180–184.

McCauley, R. N., & Bechtel, W. (2001). Explanatory pluralism and heuristic identity theory. *Theory and Psychology, 11*, 736–760.

Milner, A. D., & Goodale, M. A. (2008). Two visual systems re-viewed. *Neuropsychologia, 46*, 774–785.

Nir, Y., & Tononi, G. (2010). Dreaming and the brain: From phenomenology to neurophysiology. *Trends in Cognitive Sciences, 14*, 88–100.

Pierce, C. S., & Jastrow, J. (1884). On small differences in sensation. *Memoirs of the National Academy of Sciences, 3*, 75–83.

Prinz, J. J. (2005). A neurofunctional theory of consciousness. In A. Brook & K. Akins (Eds.), *Cognition and the brain: The philosophy and neuroscience movement*. Cambridge: Cambridge University Press.

Rosenthal, D. M. (1993). Higher-order thoughts and the appendage theory of consciousness. *Philosophical Psychology, 6*, 155–166.

Rosenthal, D. M. (2005). *Consciousness and mind*. Oxford: Clarendon.

Seth, A. K., Dienes, Z., Cleeremans, A., Overgaard, M., & Pessoa, L. (2008). Measuring consciousness: Relating behavioural and neurophysiological approaches. *Trends in Cognitive Sciences, 12*, 314–321.

Shanahan, M., & Baars, B. (2007). Global workspace theory emerges unscathed. *Behavioral and Brain Sciences, 30*, 524–525.

Sidis, B. (1898). *The psychology of suggestion; a research into the subconscious nature of man and society*. New York: D. Appleton and Company.

Skinner, B. F. (1953). *Science and human behaviour*. New York: Macmillan.

Sloman, A. (2010). An alternative to working on machine consciousness. *International Journal of Machine Consciousness, 2*, 1–18.

Sloman, A., & Chrisley, R. (2003). Virtual machines and consciousness. *Journal of Consciousness Studies, 10*, 113–172.

Snodgrass, J. M., Bernat, E., & Shevrin, H. (2004). Unconscious perception: A model-based approach to method and evidence. *Perception and Psychophysics, 66*, 846–867.

Tononi, G. (2004). An information integration theory of consciousness. *BMC Neuroscience, 5*, 42.

Tononi, G. (2008). Consciousness as integrated information: A provisional manifesto. *Biological Bulletin, 215*, 216–242.

Visser, T. A. W., & Merikle, P. M. (1999). Conscious and unconscious processes: The effects of motivation. *Consciousness and Cognition, 8*, 94–113.

Watson, J. (1913). Psychology as a behaviourist views it. *Psychological Review, 20*, 158–177.

Wilkes, K. V. (1984). Is consciousness important? *British Journal for the Philosophy of Science, 35*, 223–243.

Wilkes, K. V. (1988). Yishi, duh, um and consciousness. In A. J. Marcel & E. Bisiach (Eds.), *Consciousness in contemporary science*. Oxford: Oxford University Press.

Chapter 2
Subjective Measures of Consciousness

2.1 Introduction

This chapter forms the first part of a discussion about the problems found with a range of behavioural measures of consciousness, culminating in Chap. 4 with a methodological analysis of the very goal of identifying measures of consciousness. As the following chapters introduce much of the technical material and many of the basic ideas used throughout the book, they are necessarily more expository than later chapters. First, subjective measures based on reports are assessed here, including both the standard first order reports, as well as the use of complex second order reports used in some measures based on Signal Detection Theory. The next chapter is concerned with the problems with objective, performance-based measures of consciousness. The problems with the basic methodology used in consciousness science to identify measures of consciousness, and to distinguish conscious from unconscious perception, are developed further in Chap. 4. First, an exploration of the use of reports in consciousness sciences shows how an apparently obvious way of measuring consciousness is in fact fraught with a series of methodological problems.

Reports are still seen by many as the primary way to operationalise and measure consciousness. This is based on the assumption that subjects are reasonably (though not always) reliably informed about the contents of their experiences, and that reports accurately reflect this knowledge. This may seem like a very plausible assumption that should lead to a straightforward model of consciousness, but there can be significant discrepancies between what subjects report and how they behave. Traditionally, these discrepancies were used as support for the existence of unconscious perception, but as suggested in the introduction there is reason to query the use of reports as a way of measuring consciousness. Depending on motivation, task type, and stimulus type, subjects' reports may provide misleading evidence about what information they have concurrent access to.

This chapter is an extended version of 'New Problems with Old Measures in the Science of Consciousness' British Journal for the Philosophy of Science, 63:3, pp. 627–648, 2012. Reused with kind permission from Oxford University Press.

E. Irvine, *Consciousness as a Scientific Concept,* Studies in Brain and Mind 5,
DOI 10.1007/978-94-007-5173-6_2, © Springer Science+Business Media Dordrecht 2013

However, in recent years, the use of phenomenology and introspection in relation to subjective measures of consciousness has (again) become more popular (see e.g. Gallagher and Sorensen 2006; Lutz and Thompson 2003; Ramsøy and Overgaard 2004). The interest has spawned two special issues on 'Trusting the Subject' in the Journal of Consciousness Studies (2003, 2004 edited by Jack and Roepstorff), as well as a dedicated journal (Phenomenology and the Cognitive Sciences), which in particular has featured a special issue dedicated to 'Naturalising Phenomenology' (2004 edited by Lutz). However, little of this discussion has made direct contact with the methodological problems attributed to the use of reports outlined in psychology and psychophysics, the traditional fields in which subjective reports were used and analysed. The first part of this chapter makes this link and in doing so aims to shed some (rather old) light on the question of how reliable and useful introspective and phenomenological methods can be in a science of consciousness.

This chapter will consider three related introspective methods that have recently been proposed as a way of answering some central questions in consciousness science. Schwitzgebel (2007, 2008) suggests that the methods of immediate retrospection (reporting on experience just prior to a cue), or introspective training, may provide vital new data to help resolve questions about whether experience extends outside the bounds of attention. Somewhat differently, Ramsøy and Overgaard 2004 (Overgaard 2006; Overgaard et al. 2006) have developed a measure of consciousness called the Perceptual Awareness Scale (PAS) that assesses the graded nature of the contents of experience. As the PAS was developed with feedback from subjects, its proponents claim that it is intuitive and easy to use, and that it better reflects the graded contents of consciousness compared with traditional measures. They argue that it can therefore be used to establish the threshold between conscious and unconscious perception in a more precise way than other methods.

These introspective approaches are assessed below to see if they really can bring new evidence to discussions about the contents and boundaries of consciousness. First, it will be suggested that these approaches may not provide evidence that is clearly distinct from that provided by behavioural methods. However, the central claim is that there are significant methodological problems with all of these approaches based on the permanent presence of response bias in report, as identified by the application of Signal Detection Theory to human perception (e.g. Green and Swets 1966). While all proponents of introspective methods are aware of some potential problems with their use, (Schwitzgebel in particular is particularly skeptical), these older, and more foundational problems have been given little to no coverage in the current discussions of introspective methodology. These problems are hardly new however, this being at least the third time they have been pointed out. Instead of confirming the need to incorporate subjective reports in investigations and measures of consciousness, the continued failure of proponents of introspective measures of consciousness to address these foundational methodological problems suggests that there are ineliminable problems with their use in consciousness science.

2.2 Immediate Retrospection: Attention and Consciousness

Schwitzgebel (2007, 2008) is interested in the ongoing question in consciousness science of whether we are conscious of objects that are outside the focus of attention. Subjects are certainly unable to report about unattended stimuli in the way that they can about attended stimuli. For example, subjects report that they see all the letters in a 3-by-4 grid of letters presented for short time periods, although they are only able to identify at most four attended letters from a cued row (Sperling 1960). Subjects also report seeing a whole scene in cases of inattentional and change blindness, but fail to notice or report the changes or salient stimuli (Mack 2003; Simons and Rensink 2005). These can include failures to notice changes in the presence or absence of an engine on an aeroplane wing on consecutively shown images, and failing to notice a dancing gorilla amid a basketball game. While this behavioural evidence suggests that consciousness does not extend outside attentional focus, to many it seems intuitively plausible that we are nevertheless conscious of unattended stimuli (e.g. Block 2007; Lamme 2006; Searle 1992, 1993).

Schwitzgebel claims that arguments in favour of the rich view of conscious content, that we are conscious of unattended stimuli, are often unpersuasive. Simply stating the intuition that consciousness extends outside attention is not sufficient to show that this is the case, particularly when intuitions on this question vary. The argument that we can only notice things outside attention if we are already conscious of them is question begging about the process of attentional selection. However, Schwitzgebel also argues that the claims that consciousness does not extend outside attention cannot simply be based on behavioural evidence, such as the phenomena noted above. Whatever evidence is gathered, there are significant problems in interpreting what it means: "We already have the key data: People have some, but only a very limited, sensitivity to unattended stimuli. The question remains: Is that sensitivity (whatever it is) enough to underwrite consciousness?" (Schwitzgebel 2007, p. 12). Establishing the range of capacities that subjects have towards attended and unattended stimuli can only get us so far. What is required is a way of mapping these capacities (or the lack of them) to the presence or absence of consciousness.

Despite his 'considerable qualms' about introspection, Schwitzgebel suggests that immediate retrospection might be one way to investigate whether consciousness extends outside attention (that conscious content is 'rich') or whether it is bounded by the limits attention (that conscious content is 'thin' or 'sparse'). Immediate retrospection involves subjects reporting on the contents of experiences that have just passed. Schwitzgebel (2007) collected reports from subjects wearing a beeper that cued them to respond to a question about their experience just before the beeper went off. This is supposed to avoid reporter bias by probing subjects' experiences while they are immersed in their daily routine, rather than gathering reports in experimental situations in which subjects are likely to be concentrating on their experiences.

Schwitzgebel tested his subjects on a range of questions about their experiences, probed several times in a 3–4 hour period, including whether they were having any visual experiences at all, to whether they were having a tactile experience of their left foot. Subsequent to the beeper test, Schwitzgebel interviewed the subjects about their responses and their attitudes towards the rich and thin views of conscious content, aiming to challenge subjects and discover any obvious biases in their reports. The final responses developed from this interview process were those recorded.

The results of this experiment were very mixed, with some reports of having no visual experience at all before the beeper, but also some reports of subjects having a tactile experience of their left foot. Interestingly, responses varied both across and within subjects (for results in detail see Schwitzgebel 2007, pp. 20–22). Aside from problems in interpreting such a diverse array of reports, Schwitzgebel notes some potentially confounding factors that would affect the interpretation of any data based on immediate retrospection. These include factors that would lead to 'overreporting' the contents of experience, such as the effect that wearing a beeper might affect how subjects think about their experiences and thus how they report them, experimenter bias (Schwitzgebel holds a bias towards the view that some unattended stimuli are consciously experienced), timing errors, and confabulated reports. Factors that could lead to 'underreporting' the contents of experience include bias in subjects against using all response categories (five categories included yes/lean yes/don't know/lean no/no), failure to report 'subtle experiences', or effects of short-term memory.

The range of these potential confounds make the interpretation of subjects' reports rather difficult. In fact the question of how to map the contents of reports to the contents of consciousness seems just as complicated as the question of how behavioural evidence should be interpreted. Just as it is not obvious whether a subject's inability to 'notice' an unattended stimulus entails that they are not conscious of it, a lack of a positive report may not entail that the subjects was not conscious of, for example, their left foot before the beeper went off. The effects of practise might also mean that successful detection is not indicative of awareness, and a positive report may not indicate that subjects were conscious of what they said they were.

Just as with behavioural evidence, the contents of reports do not provide *direct* evidence about the presence or absence of consciousness in subjects, and therefore require (theoretically laden) interpretation. In this case, the problems related to the interpretation of reports means that introspective methods may not offer any methodological advantages over objective behavioural methods as a way of investigating consciousness. Schwitzgebel is however very aware of the problems raised in interpreting reports, and treats them as a way of forcing a choice in how to carry out a science of consciousness: "*If* methodological concerns in this field are inevitable, one can either be a purist and do without consciousness (or operationalize the term behaviouristically) or one can do one's best to muddle forward through doubt and ambiguity" (Schwitzgebel 2007, p. 32).

Schwitzgebel does indeed leave us with a hard decision: there are two distinct ways of investigating and measuring how subjects respond to (visual) stimuli, whose products often provide conflicting results, and neither of which provide an unproblematic inferential link to the contents of consciousness. Objective, behavioural ways of investigating consciousness don't seem to get at anything except behaviour. Subjective methods using introspective reports, such as immediate retrospection, are difficult to interpret as they may be the product of bias in the subject, the experimenter, or the task. Making inferences from behavioural responses or from verbal responses to the contents of subjects' experiences is fraught with the basic problem of how both these types of responses relate to consciousness. However, Schwitzgebel (2008, see also 2004) does offer another way of using introspection to investigate consciousness, and one that might get over some of the problems associated with immediate retrospection.

2.3 Introspective Training: The Boundaries of Consciousness

Schwitzgebel (2008) argues that naïve introspection may not provide accurate reports about the contents of experience, but that with some phenomenological training, our reports can become more reliable. Phenomenological training consists of 'attending to' and reflecting on the contents of consciousness. Most naïve reporters claim that experience consists of a fully detailed visual field, but Schwitzgebel thinks that with the right sort of training we can come to realise that this is wrong.

The training that Schwitzgebel uses is to force naïve subjects to fixate on one item in their visual field, and while fixating (and attending to other parts of their visual field), establish how much of the visual field around the fixated object actually appears 'clear'. With this training, naïve subjects come to realise that visual experiences are not as full of detailed content as originally supposed. In discussing the results of this method Schwitzgebel notes:

> Through more careful and thoughtful introspection, they seem to discover... that visual experience does not consist of a broad, stable field, flush with precise detail, hazy only at the borders. They discover that, instead, the center of clarity is tiny, shifting rapidly around a rather indistinct background. (Schwitzgebel 2008, p. 256)

Schwitzgebel's findings clearly coincide with experimental work on change and inattentional blindness that shows that subjects are unable to discriminate many objects outside the scope of spatial attention (see e.g. Mack and Rock 2003; Rensink 2005). The question is then just how different Schwitzgebel's method is from these behavioural paradigms, and thus whether it can offer new evidence about the contents of consciousness. In support of the idea that it can provide telling evidence, Schwitzgebel argues that finding out about visual acuity, for example by attempting to discriminate between a Jack or a Queen in a deck of cards presented outside foveation (2007, pp. 254–255), is not the same as investigating visual phenomenology. That is, investigating the boundaries of successful visual discrimination is not the same as investigating the 'clarity of experience'.

Yet the 'training' that Schwitzgebel uses does not significantly differ from the tasks subjects perform in change and inattentional blindness paradigms. In these paradigms subjects must either try to identify a change between two alternating (and otherwise identical) scenes, or they are tested to see whether they notice a salient feature of a scene over time. It has been found that changes are detected, and salient features are identified, only if they are saccaded or attended to (Mack and Rock 2003; Rensink 2005). Schwitzgebel's 'training' involves asking subjects to assess the effects of saccadic and attentional constraints on 'the clarity of experience'.

It is however quite unclear what 'clarity' refers to, and how subjects make judgements about the 'clarity' of an experience. It could be argued that asking if an object appears clear to me is just to ask how precisely I can discriminate it, and how confident I am in my judgments about it. That is, objects may appear clear if I can confidently detect exactly where their edges are, if I can identify the patterns on its surface, and so on. If this is the case, is not immediately obvious that trained introspective reports can add to the body of already existing behavioural evidence about the boundaries of conscious experience.

However, even if the 'clarity of experience' is a property that is distinct from a subject's discriminatory powers, there is a more serious foundational problem with the use of introspective training to uncover the contents of consciousness. This is the same problem found above with immediate retrospection; the presence of report bias. This bias can stem from the subject, from the experimenter, or from task instructions. For example, Schwitzgebel's somewhat persistent coaching style (e.g. 2007, pp. 255–256) constitutes a form of experimenter bias. Also, subjects may not be well-versed in what exactly they are supposed to be reporting in relation to visual 'clarity'. This in itself may generate the variation found in reports, including those 'trained' subjects who still insist that they 'clearly see' unattended items. As with the case of immediate retrospection discussed above, this variation in reports presents a significant problem in how they should be interpreted. Indeed, one criticism that Schwitzgebel levels at introspective techniques in general (e.g. 2004, 2008), is that they generate a wide range of reports, which contradicts the basic assumption that people experience the world in roughly the same way.

Part of this variation in reports can be attributed to the wide range of factors and processes that affect reports. Schwitzgebel (2012) has argued that introspection is not the function of a single self-monitoring system, but the product of many task-specific systems, along with central cognitive systems. For example, introspecting about visual experiences, contextual information about a scene, knowledge about your own discrimination capacities, expectations, and so on, will play a role in determining what you report: "It's more useful, I suspect, to treat the ordinary perceptual processes of vision…as part of, or as overlapping with, the introspective processes that shape my judgments about my experience" (Schwitzgebel 2012, p. 35).

Introspective training may therefore alter some of the factors that determine how judgements are made about the contents of experiences, and what introspective reports are subsequently given (e.g. by challenging expectations or contextual information). Attempting to alter these factors may improve the reliability of reports. However, it may just create one way among many of seeing and generating introspective reports. Establishing criteria by which to judge the 'correctness' of introspective reports is therefore necessary if one kind of response can be seen as better than another.

While Schwitzgebel attempts to expose obvious biases in reporting, there is in fact no independent reason to assume that extreme biases are 'bad' biases. Extreme biases may appear to generate outliers around a more 'correct' middle ground, but this does not mean that they are the product of subjects both under- and over-reporting the contents of consciousness. For example, if the thin view is the correct view about the contents of consciousness, then attempting to eliminate an extreme bias towards underreporting is in itself an experimentally induced bias. Extreme biases cannot be simply discounted as generating unacceptable variation in reports.

This ties into one way in which to judge the correctness of subjects' reports. This would be to compare subjects' reports within a group, and across different types of introspective training. Whatever is the most common type of response can form a baseline around which the contents of reports vary, with reports far from this baseline being rejected as unreliable data. Yet the evidence from Schwitzgebel's own study is highly variable, both across subjects, and across the same subject over time. There is considerable debate over the rich and thin views of conscious content precisely because people have very different and contradictory intuitions about it. In this case it is difficult to establish a norm. As Schwitzgebel himself has noted, there also seem to be cultural variations (across and within cultures over time) about the 'correct' description of experiences (e.g. Schwitzgebel 2002a; Schwitzgebel et al. 2006 on dreaming, Schwitzgebel 2002b on visual imagery, 2006 on visual phenomenology). Indeed, in the first wave of introspectionism, there was considerable disagreement across different 'schools' of training as to the properties of experience, which ultimately led to its rejection as a viable method (see e.g. Bode 1913; Boring 1953; Dodge 1912; Dunlap 1912). Historically, and currently, there are few commonly agreed on ways of 'correctly' reporting the properties of experience (also see discussion in Sect. 2.4.1 below).

This problem arises from the fact that introspection is proposed to be the *only* method of investigating conscious phenomena. Other disciplines clearly use first-person reports as evidence too, but here there are ways of normalising them. This can be done by reference to other behavioural responses and task manipulations (e.g. in psychology, where response biases in the form of demand characteristics are well-known), or by reference to some averaged response over the community in question (e.g. making grammaticality judgements in linguistics, where again the problems with the simple reliance on introspection are well known). Yet when it comes to assessing the contents of experience, these options, cross-referencing to other measures, or appealing to consensus, are not open.

First, the basic problem is that we don't know how behaviours map to conscious states. This is why we turn to subjective methods. Using behavioural responses as a way of interpreting the correctness of introspective reports is therefore not a viable option. Second, as seen above, and as explored further throughout the chapter, there is little consensus over subjects, time, cultures, (and as argued below, over of different types of training and response scales), of the 'real' properties of experience. In this case, consensus provides little guide for a baseline of correctness in reports because there *is no* consensus. The basic problem of how to establish the 'right' way of gathering reports, without reference to any other measures, is discussed in more detail below, relative to another use of introspective techniques.

2.4 Introspective Training: The Perceptual Awareness Scale

The Perceptual Awareness Scale devised by Ramsøy and Overgaard 2004 (Overgaard 2006; Overgaard et al. 2006) is a response scale based on introspective training that is meant to capture more accurate responses from subjects about the graded nature of conscious content. They claim that the graded nature of conscious experience has been persistently ignored in consciousness science, and that this has serious implications for establishing the threshold between conscious and unconscious perception. While apparently more scientifically rigorous than the introspective methods described above, this account is used to illustrate the basic problems with any use of subjective measures as ways of investigating the properties of consciousness.

One of the main targets of Ramsøy and Overgaard's criticisms is the use of dichotomous measures to indicate whether a subject is conscious of a stimulus. For example, a dichotomous measure used in identification tasks will attribute consciousness to subjects if they can correctly identify a stimulus, but not otherwise. However, subjects may still be able to tell that something was present, even if they cannot identify what it was. This means that dichotomous measures, by ignoring the graded contents of consciousness, systematically underestimate the presence of consciousness in subjects. Ramsøy and Overgaard argue that better measures and methods therefore need to be developed to record the graded nature of consciousness.

To remedy this, they suggest training subjects to define their own, graded response categories to be used in experimental situations. Developed within paradigms in which subjects had to identify several features of stimuli for a range of stimulus durations, a response scale, the Perceptual Awareness Scale (PAS), was developed with input from subjects. This scale categorises the grades of clarity with which subjects are conscious of target stimuli. The PAS consists of four response categories—'clear experience', 'almost clear experience', 'brief glimpse', and 'no experience', all matched with a verbal description of the category.

2.4.1 What Does the PAS Measure?

As Ramsøy and Overgaard note, the use of insensitive dichotomous measures of consciousness is clearly a problem in trying to establish the threshold of conscious perception. If the threshold for conscious perception is determined by the point at which subjects can no longer *identify* stimuli, then there is likely to be a great deal of conscious perception going on below this threshold. However, instead of limiting their criticisms to the use of dichotomous measures of perceptual abilities as measures of consciousness, they also criticise the use of subjective confidence ratings. Ramsøy and Overgaard argue that measures of confidence are not measures of consciousness. Instead, the relevant property to measure is the 'clarity' of experience.

Again, what exactly 'clarity' refers to is unclear. One way of examining what it means is to consider how it is interpreted by the subjects who develop the categories used in the PAS. In particular, if degrees of clarity are defined in reference to a particular set of discrimination abilities, or indeed confidence ratings, then the PAS will clearly not measure 'clarity' as outlined by Ramsøy and Overgaard. Instead, if the PAS is in fact closely related to response scales already in existence, then the PAS will not offer a new method for investigating the contents of consciousness. This idea is developed by comparing the PAS scale with an alternative framework used to characterise the graded nature of perception, not discussed by the authors.

The problem of finding a framework that captures all grades of perception, in order to find thresholds for generating different types of responses to (visual) stimuli, was largely resolved through the development Signal Detection Theory (SDT) in psychophysics beginning in the 1950's (Blackwell 1952; Eriksen 1960; Goldiamond 1958; Green and Swets 1966; Holender 1986). SDT explicitly acknowledges the graded nature of sensory discrimination, and provides a framework through which to model and measure graded perceptual abilities. These include both forced choice (objective) and 'free' (subjective report) responses concerning detection, n-alternative discrimination, categorisation, identification, and so on, as well as different types of confidence ratings. The characterization of a wide range of discrimination abilities that subjects can have towards a stimulus, and the use of graded scales to measure them, means that in contrast to the claims made by Ramsøy and Overgaard, the use of graded response scales is not new. SDT measures have been used for over half a century, in a well-established framework, and are routinely used to identify unmeasured low levels of perception in 'unconscious' perception experiments.

Before developing this comparison between SDT and the PAS it is essential to note that the SDT framework is intended to provide objective measures of perceptual abilities, while the PAS is intended to provide a measure of the subjective properties of experience. In this case, any comparison between the two may seem beside the point. However, it must be remembered that while subjective introspective measures are often contrasted with 'behavioural' measures, all reports and responses are behaviours. Whether the behaviours exhibited by subjects using introspective measures actually differ from the behaviours assessed under an objective framework is therefore a tenable question. Given that *both* subjective and objective measures are behavioural measures, a comparison between them becomes an interesting and meaningful exercise.

Given this background, it is possible to see that the graded scale of 'clarity' used in the PAS is in many ways analogous to the range of discrimination abilities and graded confidence levels identified in SDT. Evidence for this claim can be found by looking at the verbal descriptions matched with the response categories of the PAS (Overgaard et al. 2006, p. 704). For example, the category of 'no experience' is defined as 'no impression of the stimulus is experienced. All answers are experienced as mere guessing,' which points at the inability to detect a stimulus being present, combined with very low levels of confidence. The category of 'brief glimpse' is defined as 'a feeling that something was present, even though a content cannot be specified any further'. This is a prime example of a subject being able to detect but not identify a stimulus, two different types of discrimination described within SDT. The category of 'almost clear experience' is defined as 'feeling of having seen the stimulus, but being only somewhat sure about it'. This response is an example of a subject having some confidence that a specific stimulus was present, but not as much as would warrant the highest graded response. Finally, the category of 'clear experience' is defined as a 'non-ambiguous experience of the stimulus', for which subjects have high levels of confidence that a specific stimulus was present.

Interestingly, several of these verbal definitions make reference to confidence levels, despite the apparently important difference between clarity and confidence mentioned above. This suggests that the notion of clarity that Overgaard et al. appeal to is not the same as that used by subjects who use and developed the PAS. To these subjects, clarity is partly defined in terms of confidence. While certainly not conclusive, this at least shows that the notion of clarity that the PAS is supposed to assess, and the notion that it actually appears to assess, are different, and places the PAS more in line with traditional uses of confidence ratings to assess the contents of consciousness.

Aside from the similarity between the PAS and existing SDT measures, another significant problem can be found with the assumption that 'phenomenal clarity', or how clearly a stimulus is seen by a subject, is something that can provide a transferable metric across subjects, and across paradigms. This problem was commented on by earlier critics of introspective methods in psychology:

> The supposition that one experience may differ from another in 'intrinsic' clearness... results from the assumption that there is an absolute standard of what is clear... But clearness and obscureness can be construed only with reference to some specific purpose or end. Apart from such a reference the characterization has no meaning. (Bode 1913, p. 89)

The problem is that the 'clarity' of an experience may not be a property that can be evaluated in itself. Subjects' reports about how 'clearly' they see a stimulus is likely to depend on what sort of discrimination task they are asked to perform before the judgment of 'clarity' is given. In the case of the PAS, subjects are required to discriminate across a range of stimulus features (e.g. across shape, colour, location) for masked stimuli presented for a range of short durations. It has been shown that confidence ratings typically track task performance, (e.g. Haase and Fisk 2001), and it has long been acknowledged that confidence ratings also track task difficulty, with higher confidence ratings being given for easier tasks (e.g. Hertzmann 1937; Nickerson and McGoldrick 1963). Given that the verbal descriptions of the PAS refer to confidence ratings, it is likely that the way in which subjects apply the response categories of the PAS will also vary relative to task performance and task difficulty. Indeed, having carried out a comparison of different 'subjective' measures of consciousness, Sandberg et al. (2010) have found that difference subjective measures interact differently with task difficulty and performance.

This may give rise to serious problem. It is possible to make tasks easier or harder by changing non-perceptual features of task. In these cases, a stimulus is shown under identical conditions, but subjects can either be presented with two very similar responses to choose from (harder) or two very different responses (easier). Subsequent judgements of the 'phenomenal clarity' of the stimulus are likely to track the difficulty of the task. Therefore the same subject can give different PAS responses for identical stimuli, presented under identical conditions, merely by varying non-perceptual features of the task.

This point echoes that of Bode above; phenomenal clarity is always evaluated relative to a particular goal or task, so by changing the task, the judgment of clarity for an identical stimulus is likely to change. Judgements of 'phenomenal clarity' seem to track both confidence and task difficulty, despite the PAS being aimed at measuring only the phenomenal properties of consciousness. As a result, comparisons of PAS responses across different experimental paradigms are difficult to interpret, as they are always generated relative to a particular task. Overgaard et al. present the PAS as a new and precise way of gathering reports, but the way that the PAS is used and defined by subjects shows that it is at least very questionable if it succeeds in doing so.

2.4.2 Is the PAS an Exhaustive Measure?

However, there are more serious, and well-known, problems with the use of subjective reports to index awareness. One centers on the question of how to index thresholds of successful (conscious) perception. Given that the PAS can uncover more 'low level' conscious perception than traditional dichotomous subjective measures, Ramsøy and Overgaard suggest that their measure is an appropriate one to measure consciousness. In fact, the PAS is (at least in the Sandberg et al. 2010 comparative study) more 'sensitive' or exhaustive than the other graded subjective measures tested (confidence ratings and post-decision wagering). However, Ramsøy and Overgaard (2004) make a stronger, and clearly problematic claim, that the PAS is more exhaustive than objective behavioural measures of consciousness:

> As Merikle and Daneman argue, it is 'always possible to question whether any particular behavioural measure is an exhaustive measure of ALL relevant conscious experiences' (1998, p. 8). There might be significant aspects of conscious experience that are not captured by the behavioural measures. (Ramsøy and Overgaard 2004, p. 3)

Ramsøy and Overgaard present the problem of exhaustiveness as one largely related to objective behavioural measures, but it is one that is usually aimed at subjective measures. This is partly because performance thresholds given by behavioural measures are typically lower or equal to the thresholds given by subjective measures, including the PAS. Themselves opponents of purely behavioural measures, Reingold and Merikle (1990) offer a brief summary of earlier uses of introspective measures and note that subjective reports incorporate a number of biases that prevent them from being exhaustive measures of consciousness. Some of these biases stem from the instructions given to subjects (e.g. how careful subjects must be in their response, how motivated they are), or biases that come from the subject about how sure they must be before making a particular response. Reingold and Merikle state that: "[These] considerations raise serious doubts as to whether subjective reports constitute an adequate exhaustive indicator of conscious awareness... most investigators... reject any approach for distinguishing conscious from unconscious perceptual processes that is based solely on subjective reports" (Reingold and Merikle 1990, pp. 17–18).

In fact, the lack of sensitivity and exhaustiveness of subjective measures is one of the most common reasons for favouring objective over subjective measures. Objective measures based on SDT are now in standard use in subliminal perception research for precisely this reason. Indeed the main conclusion of the application of SDT to human perception is that reports are *always* biased and underestimate the abilities being tested (see Holender 1986; Reingold and Merikle 1988, 1990; Snodgrass et al. 2004). This point is extended below.

2.4.3 Response Bias

A central feature of SDT is that it identifies reports as the products of decision making about whether a particular stimulus strength warrants a particular response in a given context. According to SDT, a system will be constrained by certain objective facts about how it processes information, which determine its 'sensitivity', measured by the SDT measure d'. However, what sort of responses the system gives will also be constrained by context-dependent response criteria. For example, if the system is being rewarded for correct responses then it will try to maximize correct responses by having a very liberal response criterion. This means that the system will make positive responses even if, according to the information the system has, the probability of the stimulus being present, or being in a particular category, is very low. In contrast, if the system is punished for incorrect responses, it will try to minimize errors by having a very conservative criterion. This means that it will only make positive responses if, given the information the system has, there is a high probability that the stimulus is present, or belongs to a certain category.

The placement of response criteria is easily manipulable over subjects and over trials, and is sensitive to factors such as task type, task instructions, and motivation. These and other factors are noted by Schwitzgebel in his discussion of the problems involved in interpreting reports. What SDT does is to formalize these problems; objective measures such as d' measure a system's objective sensitivity to stimuli, while the contents of reports are governed both by d' and by internally set response criteria.

The existence of response criteria shows that it is not obvious what a most 'natural' or 'correct' report of an experience would be. Bayne and Spener (2010) state that "... we should not think of introspection as invariably subject to the influence of expectation" (p. 17). Based on this they suggest that 'trustworthy' introspective reports can be identified as those for which expectations and response biases are largely absent. However, this is just to misunderstand how reports are generated. Expectations and response bias are always present, as they are of fundamental importance to the decision-making that is part of report generation. Thus, Snodgrass and Lepisto (2007) state that reports are *never* free from experimentally induced or contextually driven bias: "... contrary to many researchers' implicit assumptions, there is no such thing as an unmediated 'subjective report'—ever" (p. 526).

This means that it is difficult to argue that introspective training can produce more or less 'correct' reports about the contents of experience. Instead, subjects who have undergone phenomenological training are equipped with different response biases to the ones they started with. A focus on training subjects to report 'low level' experiences that they might otherwise discount (as with the PAS) ensures that they lower their response criterion. However, their response criterion is not eradicated, and still biases reports, just in a different way. Any type of training that succeeded in eradicating response bias would simply replicate the objective measure d', and would thus defeat the purpose of using introspective methods to provide a subjective measure.

The persistent and ineradicable presence of bias in subjective measures not only affects the PAS, but all introspective methods (including those explored by Schwitzgebel, as he notes). Combined with earlier arguments, it becomes apparent that not only will the metric of introspective measures be specific to particular tasks, but that they will be affected by a whole range of other non-perceptual factors, including motivation, training and the goal of a subject. It could be argued that this is not necessarily a problem for introspective measures of experience; that it is likely that these sorts of measures would be context-specific. Yet given the range of biases that subjects might bring to a task, there is still the open question of which type or degree of bias is the one that produces the most veridical reports about the contents of experience (the aim of introspective methods in general).

However, coming back to an earlier problem, there is no principled way of labeling response biases as universally 'good' or 'bad'. Biases can be identified as more or less appropriate according to the goals of a subject (e.g. to minimize errors), but the same bias may be totally inappropriate for achieving other goals (e.g. to maximize hits). Biases may be developed during training such that they fit the expectations of an experimenter or a particular introspective framework, but different training may produce a very different set of biases, and thus a very different set of responses. Which set of biases and reports are more 'correct' in this case is not obvious, other than that these various induced biases are appropriate to the context they are used in. Sometimes response biases better reflect the real limits on our discrimination capacities (e.g. that discrimination abilities are minimal outside fixation, or that successful detection can occur with very low stimulus strength), but this merely makes reports more accurate relative to objective discrimination abilities.

So, response biases can either be judged more or less appropriate to a particular context or goal, or they can be judged relative to the baseline provided by objective measures of task performance. As noted earlier, the second option is less likely to fit with the aims of those proposing introspective measures; subjective measures are intended to measure fundamentally different states to objective measures. Curiously however, objective measures are used to provide a comparison and background metric for subjective measures, particularly in order to evaluate the relative exhaustiveness of different subjective measures (e.g. Sandberg et al. 2010). If, as these authors claim, the best measure is the one that best tracks objective measures of task performance, it is not clear why objective measures should not be used instead (as they are by this definition maximally exhaustive).

However, the first option, that introspective responses are recognized as being context-sensitive, and can only be evaluated as better or worse in relation to a task or goal, also seems a problematic option. Aside from going against the general assumption in the introspection literature that there are phenomenal facts, and that introspective methods can be used to uncover them, it becomes nonsensical to talk about the 'best' way of training subjects to report about their experiences. This is because the goal of satisfying the experimenter, (e.g. by following instructions and lowering your response criterion), would ensure that all forms of bias as induced by different training techniques would lead to 'correct' reports. Once it is accepted that eradicating bias is impossible in a subjective measure, many types of bias, and thus many types of report, become viable and appropriate ways of responding to the same stimulus (given an appropriate goal or context).

It is hard to see how this extreme relativism could produce a practical, useful measure. Indeed, the context-sensitivity and variability of subjective measures was used as a dominant argument against the use of subjective measures. Given that introspective methods are often put forward as ways of settling long-term disputes in consciousness science that presuppose some roughly invariant answer (the nature of visual experience outside attention, the threshold for unconscious perception), the inherent variability in subjective measures makes it difficult to see how they could settle these questions.

In summary, either the 'correctness' of introspective reports are evaluated in relation to objective measures, or the appropriateness of response bias is evaluated in relation to the achievement of a context-specific goal or task. Neither option preserves the central motivation behind the use of introspective methods; that there should be a way of providing a practical, context-independent measure of the subjective contents of experience that differs from existing objective measures.

2.4.4 Natural Response Categories?

Further, there are reasons to believe that both the number and the descriptors given to the response categories that subjects form in the PAS are subject to non-perceptual constraints. Overgaard et al. (2006) claim that the PAS is an 'intuitive' measure that is easy to use, and that the practice of getting subjects to form their own response categories creates a 'natural' response scale that better reflects the real graded quality of experience. However, a graded response scale is constrained by response criteria in the same way that a dichotomous scale is. Dichotomous response scales force subjects to generate only one response criterion (information threshold or probability) over which a certain type of response (e.g. 'stimulus present') is appropriate. Graded response scales force subjects to generate multiple response criteria for different kinds of responses, from 'clear experience' (a lot of stimulus information) to 'brief glimpse' (less stimulus information) to 'no experience' (very low levels of stimulus information). Different levels of stimulus information across different tasks can all be categorized as instances of 'brief glimpses', depending on how liberal or conservative a response criterion is for that category, for a particular subject.

In support of the idea that four response categories are the most 'natural' number to use, Overgaard et al. note that in the Ramsøy and Overgaard (2004) study, "When subjects tried to use more than four categories in the scale, they found it confusing and quickly abandoned the extra categories" (Overgaard et al. 2006, p. 702). Presumably there are more than four ways of responding to stimuli, but in experimental paradigms where subjects are also tested on a range of visual discrimination tasks (as in the PAS), subjects clearly find that keeping track of more than four categories is difficult, possibly due to working memory constraints. The PAS must be useable by subjects, so is formed around the constraints of the task. Perhaps for more simple tasks, more response categories could be used.

Conversely, there presumably are situations (such as the attentional blink discussed in Overgaard et al. 2006), for which dichotomous responses are easier to generate. In Sergent and Dehaene's (2004) attentional blink paradigm, subjects are given a 20 point visibility scale with only the extreme points given a verbal description. They found that subjects tended to use the extreme points on the scale far more than the points in the middle. Overgaard et al. note that as different points in the scale were inadequately described, subjects naturally tended to use on the extreme ends of the scale in their responses.

Under some experimental constraints, with a large number of undefined response categories, subjects find it easier and more 'natural' to respond in a dichotomous way. When instructed to form their own graded response categories the easiest number of categories to use is four. It is true that using a continuous scale in a dichotomous way may be the artificial result of non-perceptual factors, such as the scale being undefined. However, the use of four response categories is also an artificial result of non-perceptual factors, such as experimenter's instructors to use a non-dichotomous scale, and the limits of working memory. The PAS is typically used in very similar paradigms, which means that its intuitive appeal and ease of use have not been tested outside these limited conditions (n-alternative feature discrimination followed by PAS response, see Overgaard et al. 2006; Ramsøy and Overgaard 2004; Sandberg et al. 2010). In an easier task, or with the right balance of rewards and costs, it is plausible that a larger number of response categories would be the most natural one to use. As a measure that is supposed to reflect the general nature of the contents of consciousness, the PAS may in fact be the product of a set of very paradigm-specific constraints.

2.5 Summary: Subjective Measures and Introspective Approaches

Three different ways of attempting to train subjects to produce accurate and reliable introspective reports about the contents of consciousness have been discussed. Schwitzgebel offers immediate retrospection and introspective training as ways of getting accurate introspective reports. As he acknowledges, his results are mixed, and are plausibly influenced by response biases, so are difficult to interpret. Further, there appears to be no non-circular way of establishing which are 'good' and 'bad' biases in reports, so these introspective techniques do not offer a solution to the problem of how rich the contents of consciousness really are.

The Perceptual Awareness Scale proposed by Ramsøy and Overgaard was also argued to suffer from similar problems. This response scale was developed with subjects in order to provide an intuitive and 'natural' scale with which to categorise the graded clarity of the contents of consciousness. The notion of 'phenomenal clarity' that the PAS is intended to measure is distinct from visual discrimination abilities and confidence ratings, and can be compared across different contexts. However, as evidenced by the way that the PAS is used and interpreted by subjects, the PAS fails to capture such a notion. Further, by comparing the PAS to the framework provided by Signal Detection Theory, it was argued that the PAS was neither new, exhaustive, unbiased nor 'natural'.

SDT shows that in contrast with the assumptions held by Schwitzgebel and the proponents of the PAS (and likely the proponents of similar approaches), all responses are biased, and that this is a central feature of report generation. To repeat Snodgrass and Lepisto (2007) "there is no such thing as an unmediated 'subjective report'—ever" (p. 526). While there may be more or less appropriate response biases for achieving different goals in certain contexts, none of these response biases produce more or less 'correct' reports, and they cannot be generalized to reflect a general tendency to respond to stimuli in a certain way. This is because there is no context-independent most 'natural' way to respond to a stimulus. Providing a set of phenomenological training to generate 'accurate' phenomenological reports, or providing a subjective response scale with which to categorise the graded nature of phenomenal 'clarity' are ill-formed goals.

The problems with report-based measures of consciousness have long been known and are again coming to light with the reintroduction of introspection-based measures. The recognition that reports are never free from bias must be made again, and the issues raised by the application of SDT to human perception must go through another round of reiteration. In order to validate the use of introspective or subjective measures in consciousness science the problems raised above need to be dealt with. The following section appraises an alternative approach that tries to deal with the problem of response bias, yet also preserve the use of subjective measures of consciousness.

2.6 Type 2 Confidence Ratings as a Measure of Consciousness

Given the rejection of phenomenologically informed measures above, it might appear as though any report-based, subjective measure will be severely methodologically flawed. There are attempts however to combine the intuitive appeal of report-based measures and the mathematical rigour offered by objective measures of consciousness. This section builds on a suggestion from Frith and Lau (2006) to apply Signal Detection Theory to internal states, which could provide a methodologically sound, and subjective, measure of consciousness. Within this framework, questions about

Table 2.1 Expected results from subjects who are conscious of a stimulus

	Type 2 high confidence	Type 2 low confidence
Correct response to Type 1 task	✓	×
Incorrect response to Type 1 task	×	✓

the contents of consciousness are somewhat sidelined, with the emphasis instead on finding a measure that can be used to clearly distinguish conscious from unconscious perception. This section is therefore more directly aimed at one of the questions that Ramsøy and Overgaard are interested in; how to find a measure of consciousness that can be used to investigate the limits of conscious perception, and by implication the breadth of unconscious perception.

Two different measures can be outlined using Signal Detection Theory; one is how sensitive subjects are to external stimuli (Type 1 sensitivity d' discussed above), while the other assesses how well subjects can judge whether they have made correct or incorrect Type 1 responses (Type 2 measure a') (for original use of Type 2 confidence ratings see Clarke et al. 1959; Pollack 1959). That is, Type 2 measures assess how well subjects can monitor their own internal information and make judgements about its reliability.

As an index of responses that subjects make to internal information, rather than responses made to external stimuli only, Type 2 measures seem more intuitively linked with the subjective aspect of visual consciousness than Type 1 measures. Combined with the methodological advantages of the SDT framework, including the ability to generate bias free measures, this makes Type 2 measures an alternative candidate measure of consciousness. Thus, Kunimoto et al. (2001) suggest that Type 2 confidence ratings provide a bias free and phenomenologically valid (i.e. intuitively plausible) measure of awareness. Lau's (2008) Higher Order Bayesian decision theory of consciousness also suggests that the ability to adequately model the accuracy of one's own performance, as measured through Type 2 confidence ratings, is integral to consciousness.

More precisely, these proposals are based on the reasoning that if subjects are aware of stimuli, then they should be confident about their responses of whether targets are present or not. If subjects are conscious of target stimuli, high Type 2 confidence ratings should correlate with correct Type 1 responses (those that the subject is sure they got right), and low Type 2 confidence ratings should correlate with incorrect Type 1 responses (those that they are less sure about). On the other hand, if subjects are not aware of targets, then they will not know whether their responses are correct or not, so there should be no relationship between Type 1 responses and Type 2 confidence ratings. Kunimoto et al. define a measure of consciousness, a', based on these relationships, summarised below and in Table 2.1:

> ... [subjects'] awareness can be assessed by the relationship between their confidence judgments and accuracy, under the plausible assumption that their confidence cannot reflect their accuracy unless they are at least partially aware of the information on which they based their discriminative responses. (Kunimoto et al. 2001, p. 302)

2.6.1 Comparisons Between d' and a'

One way in which the measure a' has been used to index consciousness is to compare it with the measure d'. For example, blindsight subjects such as GY perform well on Type 1 discriminatory tasks (high d'), but fail to consistently identify which are their own correct and incorrect responses in a 'commentary key paradigm' or post-task wager (both Type 2 tasks; low a') (Persaud et al. 2007; Weiskrantz 1998). This has been taken as evidence that GY is not aware of stimuli in his blind field: "That GY was capable of using visual information in his scotoma to perform the discrimination, yet did not maximize his winnings by consistently wagering high after correct classifications, indicates that he was not always aware that he was making correct decisions" (Persaud et al. 2007, p. 258).

In other applications of this method of comparing d' and a', Szczepanowski and Pessoa (2007) reported a dissociation between Type 1 and Type 2 task performance for fearful-face perception, concluding that the two tasks index qualitatively different processes. Lau and Passingham (2006) also showed that Type 2 task performance can differ in two performance-matched cases of a Type 1 task, which they take as evidence of 'relative blindsight' or unconscious perception.

However, a direct comparison between Type 1 and Type 2 performance may be misleading, and does not necessarily indicate a lack of awareness. The reasons behind this also suggest that the use of a' to index consciousness may be fundamentally problematic. To show this, Galvin et al. (2003, pp. 849–854) provide an example of a task in which Type 1 performance is better than Type 2 performance, despite a subject being fully conscious of the stimuli used. In this task, a subject is asked to report the number showing on one die (that shows only '0's and '3's), given the sum of this number with the numbers from two normal die (i.e. that carry the numbers 1–6, one number on each face). Three stipulations are made in this example: (1) all available information is used to generate responses in both Type 1 and Type 2 tasks, (2) optimal response strategies are used in both tasks, and (3) the subject is *fully aware* of the stimulus in both tasks.

For example, if the numbers showing on all three dice sum to 12, the subject must report whether the abnormal die shows a '0' or a '3', given probabilistic information about either situation being the case. As there are more ways of making 9 with two normal dice (3–6, 4–5, 5–4, 6–3) than a 12 (6–6), the most likely number on the abnormal die is a '3', so this is what the subject should report for the Type 1 task. The subject can also give a 'high confidence' response in the Type 2 task. However, due to the statistical features of the example, sometimes the subject will report the correct answer in the Type 1 task, but should give a 'low confidence' response in the Type 2 task, even though all information is used optimally in the calculation (see Appendix 1 for further details).

In this example, Galvin et al. show that Type 1 performance can outstrip Type 2 performance even when all stimulus information is fully accessible, conscious, and optimally used by a subject, due to statistical features of the task. Galvin et al. identify the factors that can lead to greater Type 1 performance compared with Type 2 performance:

> The relationship between Type 1 and Type 2 discriminations depends on the performance measure chosen, the decision axes chosen for each of the two tasks, the Type 1 criterion used, the shape of distributions underlying the Type 1 decision, and the prior probabilities of the Type 1 events. (Galvin et al. 2003, p. 860)

Differences between Type 1 and Type 2 task performance cannot therefore be taken as simple indications of the presence or absence of consciousness. Type 1 performance may be greater than Type 2 performance not because of a lack of consciousness, but because of any of the factors that Galvin et al. describe. Those who have used comparisons between Type 1 and Type 2 tasks to measure consciousness (such as Weiskrantz, Szeczepanowksi, Persaud) must therefore show more decisively that the difference they found really is evidence of a lack of consciousness rather than the product of statistical features of the experimental paradigm they used.

Lau (2008; Lau and Passingham 2006) has tried to control for at least some of these factors in his use of comparisons between Type 1 and Type 2 performance to investigate the presence or absence of consciousness. For example, Lau and Passingham (2006) control for performance in their experimental design by making use of two different stimulus conditions that give rise to the same Type 1 performance in meta-contrast masking. For very short and very long time periods (stimulus onset asynchrony, or SOAs) between the target and the meta-contrast mask, Type 1 performance is roughly the same, but for a certain band of SOAs performance suffers, creating a U-shaped performance distribution. By matching performance levels on either side of the 'U', Lau and Passingham were able to test Type 2 performance on two different stimuli that subjects had been equally successful at discriminating in the Type 1 task. They found that Type 2 performance was much lower for short SOAs than long SOAs, and used this as evidence that subjects were not (as) aware of stimuli presented with short SOAs as long SOAs. They claim that performance matching is a novel and productive way of comparing Type 1 and Type 2 performance, and thereby measuring consciousness.

However, this method of matching performance levels only targets some of the factors that determine Type 1 and Type 2 task performance described by Galvin et al. The decision axes, Type 1 criteria or distributions of the Type 1 decisions could be different on either side of the 'U'. In this case, Type 2 responses could be generated differently for short and long SOAs. For example, subjects may be more conservative in their responses to stimuli with masks presented for short SOAs (a difference in Type 1 criteria), or the asymmetry of the decision distributions on either side of the 'dip' (tracking performance as it decreases then increases again) could affect Type 2 performance. In itself this could determine the relationship between Type 1 and Type 2 task performance in such a way that it need not reflect the presence or absence of consciousness. What Lau and Passingham need to show more clearly is that the differences in Type 2 performance are not an artifact of the experiment, due to the different ways that subjects make decisions over the different SOAs.

2.6.2 Comparing Thresholds for d' and a'

A way to use Type 2 performance as a measure of consciousness that is not subject to the problems above is to compare thresholds for above chance performance in Type 1 and Type 2 tasks. It has been argued that if the conditions for which the measure of awareness is zero ($a' = 0$) are different to the conditions under which the measure of sensitivity is zero ($d' = 0$), then the presence of unconscious perception can inferred. That is, subjects would still be able to complete the Type 1 task above chance level, but fail to monitor their responses in such a way to perform well at the Type 2 task. This approach was used by Kunimoto et al. (2001) who claim to have found evidence of unconscious perception in a small but significant difference (3ms SOA) between objective ($d' = 0$) and subjective ($a' = 0$) thresholds in a visual discrimination task.

However, the very finding of this difference in thresholds provides questionable evidence for the presence of unconscious perception. Galvin et al. (2003) suggest that it should be *impossible* to find different thresholds for Type 1 and Type 2 task performance if subjects are doing the Type 2 task as it is meant to be done. That is, if subjects rely on different sources of information for the two tasks, the Type 2 task no longer functions as a subjective commentary on first order task. In this case, the Type 2 rating cannot be used to measure awareness for the Type 1 task. The fact that different thresholds for a' and d' were found by Kunimoto et al. suggest "that the form of the instruction is very important to the [subject] and that being asked to discriminate and to give a commentary on one's performance causes one to rely on different sources of information" (Galvin et al. 2003, p. 861).

2.6.3 Bias and Phenomenological Validity

Along with the problems inherent in simple comparisons between Type 1 and Type 2 performance, and comparisons of thresholds for a' and d', there is also the possibility that a' fails to be free from response bias. Additionally it is possible that the measure does not fulfill Kunimoto et al.'s other condition that a measure must be phenomenologically valid; i.e. that it make sense on intuitive grounds. If the measure fails on any of these grounds, then it does not fulfill the necessary criteria to function as an adequate measure of awareness.

The advantage of using SDT analysis on Type 2 responses to give a' is that it is supposed to give a measure of awareness that is not affected by the response bias of subjects. However, Evans and Azzopardi (2007) have shown that a' varies with experimentally induced response bias. In particular, they showed that a' for a blindsight subject varied through a range of positive values without approaching zero, which is inconsistent with other measures of awareness in blindsight. They point out that response bias was artificially clamped in Kunimoto et al.'s (2001) study by limiting the numbers of each type of response subjects could give. In this case, a' clearly fails to be a bias free measure of Type 2 confidence ratings, and so according to the equation of Type 2 task performance with awareness, a' also fails to be a bias free measure of awareness.

As a way of overcoming many of the problems associated with the use of a', Rounis et al. (2010) have developed an alternative measure of Type 2 performance, meta-d' . Meta-d' is an estimate of how much information subjects use to generate Type 2 responses, compared with how much they use for Type 1 responses. Meta-d' is a measure of what d' an ideal observer would have (i.e. using all available information to generate responses) in order to generate the observed Type 2 responses of a subject, given estimated values of their criterion level β. In ideal observers, meta-$d' = d'$ because ideal observers use the same sets of information for both Type 1 and Type 2 tasks. For those whose Type 2 ratings totally fail to track Type 1 performance, meta-$d' = 0$ because they do not make use of any information to generate their Type 2 responses (guessing). For normal subjects, who are somewhere in between (though see response from Galvin et al. above), meta-$d' < d'$, as they make use of less information for the Type 2 than the Type 1 task. Given an equation between Type 2 decision-making and consciousness, meta-d' could provide a measure of consciousness, as it tracks Type 2 decision-making in an unbiased way.

However, even if meta-d' is a bias free measure of metacognitive sensitivity, there are significant concerns about the equation between second order confidence ratings used in Type 2 tasks and consciousness. In response to claims that post-decision wagering (see Persaud et al. 2007), a form of Type 2 confidence rating task, 'directly' measures awareness, Seth (2008) argues that meta-cognitive content is not the same as conscious content. That is, measures of Type 2 performance are not measures of consciousness.

Whether Type 2 confidence ratings are phenomenologically valid measures of consciousness is discussed in more detail by Lau in his Higher Order Bayesian decision theory of consciousness (2008). In this, he argues that while d' is simply a measure of information processing, the setting and maintaining of response criteria measured by Type 2 confidence ratings is relevant to consciousness. In this case, while a' and meta-d' measure second order decision-making, this is equivalent to a basic level of perceptual consciousness, as found in Higher Order Thought theories of consciousness. However, Lau suggests only a 'minimal interpretation' of his model, and claims that "It is not supposed to explain all features of consciousness" (Lau 2008, p. 46).

Although the use of confidence based measures of awareness is reasonably popular, there has been little argument to justify the link between confidence ratings and consciousness. Kunimoto et al.'s (2001) argument for the use of confidence ratings relies on the 'plausible' assumption that a correlation between confidence and discriminative accuracy cannot occur without subjects being aware of the information on which they base their discriminative responses. While this assumption may seem reasonable for some, on a purely information processing account there is no reason why accuracy and confidence must be mediated by consciousness. It may therefore be possible to generate confidence responses in the absence of consciousness (see Koriat 2007).

While Kunimoto et al. acknowledge this problem, they suggest that this sort of argument is valid against *any* behavioural measure of awareness. According to Kunimoto et al. (and possibly Lau) it seems that given the other options, confidence ratings just appear more likely to index awareness. However, this reasoning is far from sufficient to establish that a' or meta-d' is more likely than any other measure to exclusively measure awareness. The general problems in establishing *any* measure as a measure of consciousness are explored in the next two chapters, where it is argued that aside from intuitively 'plausible' assumptions there are no methodological reasons to use one type of measure over another as a measure of consciousness.

2.7 Conclusion

Two different ways of generating an adequate subjective measure of consciousness have been discussed, and in both cases it has been argued that there are significant problems with their application. Phenomenological training, an attempt to get more accurate reports in the form of Type 1 responses from subjects, suffers the same problems as earlier attempts to use introspective reports to investigate consciousness. Both Schwitzgebel's introspective training and Overgaard's Perceptual Awareness Scale suffer from the essential problem that reports are always subject to response bias, and that there are no principled ways of dealing with this bias in consciousness research. Importantly, no arguments are offered to overcome the serious methodological problems with subjective measures of consciousness developed since the 1950s. In this case modern proponents of using (Type 1) reportability as a measure of consciousness cannot justify the reintroduction of introspective methods and measures.

It has also been argued that the attempt to combine the intuitive advantages of subjective reports, and the mathematical framework of SDT, using Type 2 confidence ratings also fails to provide an adequate measure of consciousness. Neither comparisons between d', a' or meta-d', nor comparisons of thresholds for d' and a' or meta-d', can be used to infer the presence or absence of consciousness. The measure a' also fails to be bias free, and while bias free measures of Type 2 sensitivity exist (meta-d'), there are reasons to doubt the phenomenological validity of Type 2 measures in general.

While subjective measures of consciousness seem the most intuitively plausible measures that can and should be used in consciousness science, the methodological problems outlined above show how they cannot simply be assumed to be the right ones. In fact, both objective and subjective measures can be easily translated as measures of different cognitive capacities, (sensitivity and decision-making), in which case the question arises of which, if any, cognitive capacity can be identified with consciousness. The problems related to the definition of consciousness, and the assumption that it is a phenomenon that exists beyond the many specific ways in which it is operationalised, return throughout later chapters. In contrast with the focus on subjective measures of consciousness found in this chapter, the next chapters address the problems associated with objective measures of consciousness, exemplified by d', and whether there are any methodological solutions to the general problem of how to identify an appropriate measure of consciousness.

References

Bayne, T., & Spener, M. (2010). Introspective humility. *Philosophical Issues, 20,* 1–22.

Blackwell, H. R. (1952). Studies of psychophysical methods for measuring visual thresholds. *Journal of the Optical Society of America, 42,* 606–616.

Block, N. (2007). Consciousness, accessibility, and the mesh between psychology and neuroscience. *Behavioral and Brain Sciences, 30,* 481–548 (inc. responses).

Bode, B. H. (1913). The method of introspection. *The Journal of Philosophy, Psychology and Scientific Methods, 10,* 85–91.

Boring, E. G. (1953). A history of introspection. *Psychological Bulletin, 50,* 169–189.

Clarke, F. R., Birdsall, T. G., & Tanner, W. P., Jr. (1959). Two types of ROC curves and definitions of parameters. *Journal of the Acoustical Society of America, 31,* 629–630.

Dodge, R. (1912). The theory and limitations of introspection. *The American Journal of Psychology, 23,* 214–229.

Dunlap, K. (1912). The case against instrospection. *Psychological Review, 19,* 404–413.

Eriksen, C. W. (1960). Discrimination and learning without awareness: A methodological survey and evaluation. *Psychological Review, 67,* 279–300.

Evans, S., & Azzopardi, P. (2007). Evaluation of a 'bias-free' measure of awareness. *Spatial Vision, 20,* 61–77.

Frith, C. D., & Lau, H. C. (2006). The problem of introspection. *Consciousness and Cognition, 15,* 761–764.

Gallagher, S., & Sorensen, J. B. (2006). Experimenting with phenomenology. *Consciousness and Cognition, 15,* 119–134.

Galvin, S. J., Podd, J. V., Drga, V., & Whitmore, J. (2003). Type 2 tasks in the theory of signal detectability: Discrimination between correct and incorrect decisions. *Psychonomic Bulletin and Review, 10,* 843–876.

Goldiamond, I. (1958). Indicators of perception: 1. Subliminal perception, subception, unconscious perception: An analysis in terms of psychophysical indicator methodology. *Psychological Bulletin, 55,* 373–411.

Green, D. M., & Swets, J. A. (1966). *Signal detection theory and psychophysics*. New York: Wiley.

Haase, S. J., & Fisk, G. (2001). Confidence in word detection predicts word identification: Implications for an unconscious perception paradigm. *The American Journal of Psychology, 114,* 439–468.

Hertzmann, M. (1937). Confidence ratings as an index of task difficulty. *Journal of Experimental Psychology, 21,* 113–119.

Holender, D. (1986). Semantic activation without conscious identification in dichotic listening, parafoveal vision, and visual masking: A survey and appraisal. *Behavioral and Brain Sciences, 9,* 1–23.

Jack, A. I., & Roepstorff, A. (2003). Trusting the Subject I. Special issue of *Journal of Consciousness Studies, 10,* 9–10.

Jack, A. I., & Roepstorff, A. (2004). Trusting the Subject II. Special issue of *Journal of Consciousness Studies, 11,* 7–8.

Koriat, A. (2007). Metacognition and consciousness. In D. Zelazo, M. Moscovitch, & E. Thompson (Eds.), *The Cambridge handbook of consciousness* (pp. 289–325). Cambridge: Cambridge University Press.

Kunimoto, C., Miller, J., & Pashler, H. (2001). Confidence and accuracy of near-threshold discrimination responses. *Consciousness and Cognition, 10,* 294–340.

Lamme, V. A. F. (2006). Towards a true neural stance on consciousness. *Trends in Cognitive Sciences, 10,* 494–501.

Lau, H. C. (2008). A higher order Bayesian decision theory of consciousness. In R. Banerjee & B. K. Chakrabati (Eds.), *Progress in brain research* (Vol. 168, pp. 35–48). Amsterdam: Elsevier.

Lau, H. C., & Passingham, R. E. (2006). Relative blindsight in normal observers and the neural correlate of visual consciousness. *Proceedings of the National Academy of Sciences, 103,* 18763–18768.

Lutz, A. (Ed.) (2004). Special Issue: Naturalising Phenomenology. *Phenomenology and Cognitive Science, 3*:4.

Lutz, A., & Thompson, E. (2003). Neurophenomenology: Integrating subjective experience and brain dynamics in the neuroscience of consciousness. *Journal of Consciousness Studies, 10,* 31–52.

Mack, A. (2003). Inattentional blindness: Looking without seeing. *Current Directions in Psychological Science, 12,* 180–184.

Mack, A., & Rock, I. (2003). Inattentional blindness: A review. *Directions in Psychological Science, 12,* 180–184.

Merikle, P. M., & Daneman, M. (1998). Psychological investigations of unconscious perception. *Journal of Consciousness Studies, 5,* 5–18.

Nickerson, R. S., & McGoldrick, C. C., Jr. (1963). Confidence, correctness, and difficulty with non-psychophysical comparative judgments. *Perceptual and Motor Skills, 17,* 159–167.

Overgaard, M. (2006). Introspection in science. *Consciousness and Cognition, 15,* 629–633.

Overgaard, M., Rote, J., Mouridsen, K., & Ramsøy, T. Z. (2006). Is conscious perception gradual or dichotomous? A comparison of report methodologies during a visual task. *Consciousness and Cognition, 15,* 700–708.

Persaud, N., McLeod, P., & Cowey, A. (2007). Post-decision wagering objectively measures awareness. *Nature Neuroscience, 10,* 257–261.

Pollack, I. (1959). On indices of signal and response discriminability. *Journal of the Acoustical Society of America, 31,* 1031.

Ramsøy, T. Z., & Overgaard, M. (2004). Introspection and subliminal perception. *Phenomenology and the Cognitive Sciences, 3,* 1–23.

Reingold, E. M., & Merikle, P. M. (1988). Using direct and indirect measure to study perception without awareness. *Perception and Psychophysics, 44,* 563–575.

Reingold, E. M., & Merikle, P. M. (1990). On the inter-relatedness of theory and measurement in the study of unconscious processes. *Mind and Language, 5,* 9–28.

Rensink, R. A. (2005). Change blindness: Implications for the nature of visual attention. *Trends in Cognitive Sciences, 9,* 16–20.

Rounis, E., Maniscalco, B., Rothwell, J. C., Passingham, R. E., & Lau, H. (2010). Theta-burst transcranial magnetic stimulation to the prefrontal cortex impairs metacognitive visual awareness. *Cognitive Neuroscience, 1,* 165–175.

Sandberg, K., Timmermans, B., Overgaard, M., & Cleeremans, A. (2010). Measuring consciousness: Is one measure better than the other? *Consciousness and Cognition, 19,* 1069–1078.

Schwitzgebel, E. (2002a). Why did we think we dreamed in black and white? *Studies in the History and Philosophy of Science, 33,* 649–660.

Schwitzgebel, E. (2002b). How well do we know our own conscious experience? The case of visual imagery. *Journal of Consciousness Studies, 9,* 35–53.

Schwitzgebel, E. (2004). Introspective training apprehensively defended: Reflections on Titchener's lab manual. *Journal of Consciousness Studies, 11,* 58–76.

Schwitzgebel, E. (2006). Do things look flat? *Philosophy and Phenomenological Research, 72,* 589–599.

Schwitzgebel, E. (2007). Do you have constant tactile experience of your feet in your shoes? Or is experience limited to what's in attention? *Journal of Consciousness Studies, 14,* 5–35.

Schwitzgebel, E. (2008). The unreliability of naïve introspection. *Philosophical Review, 117,* 245–273.

Schwitzgebel, E. (2012). Introspection, what? In D. Smithies & D. Stoljar (Eds.), *Introspection and consciousness* (pp. 29–48). Oxford: OUP.

Schwitzgebel, E., Huang, C., & Zhou, Y. (2006). Do we dream in color? Cultural variations and skepticism. *Dreaming, 16,* 36–42.

Searle, J. R. (1992). *The rediscovery of the mind.* Cambridge: MIT.

Searle, J. R. (1993). The problem of consciousness. *Consciousness & Cognition, 2,* 310–319.

Sergent, C., & Dehaene, S. (2004). Is consciousness a gradual phenomenon? Evidence for an all-or-none bifurcation during the attentional blink. *Psychological Science, 15,* 720–728.

Seth, A. K. (2008). Post-decision wagering measures metacognitive content, not sensory consciousness. *Consciousness and Cognition, 17,* 981–983.

Simons, D. J., & Rensink, R. A. (2005). Change blindness: Past, present, and future. *Trends in Cognitive Sciences, 9,* 16–20.

Snodgrass, J. M., & Lepisto, S. A. (2007). Access for what? Reflective consciousness. *Behavioral and Brain Sciences, 30,* 525–526.

Snodgrass, J. M., Bernat, E., & Shevrin, H. (2004). Unconscious perception: A model-based approach to method and evidence. *Perception and Psychophysics, 66,* 846–867.

Sperling, G. (1960). The information available in brief visual presentations. *Psychological Monographs: General and Applied, 74,* 1–29.

Szczepanowski, R., & Pessoa, L. (2007). Fear perception: Can objective and subjective awareness measures be dissociated? *Journal of Vision, 7,* 1–17.

Weiskrantz, L. (1998). Consciousness and commentaries. *International Journal of Psychology, 33,* 227–233.

Chapter 3
Measures of Consciousness and the Method of Qualitative Differences

3.1 Introduction

Following on from the discussion of subjective measures in Chap. 2, this chapter begins with an investigation of the use of objective measures in consciousness research. The objective measure d' is analysed by considering arguments in favour of its use, based on the existence of response bias, and arguments against its use, including those based on the phenomenon of blindsight. Both subjective and objective measures are also assessed through the method of treating consciousness as the exhibition of control.

Given the problems that have already been identified with subjective measures of consciousness, and those discussed below with the use of objective measures, the debate between proponents of these two types of measures will be explored. As a way of resolving the debate, experimental methods or theoretical arguments have been sought that would decisively favour using one type of measure. One method is that of identifying qualitative differences in behaviour, and using these to establish a measure of consciousness. It has been suggested that the differences in behaviour that distinguish between conscious and unconscious perception are likely to be qualitative, rather than quantitative, in character. Qualitative differences mark discrete differences in behaviour, for example the ability to perform a task compared with the inability to perform it. These contrast with quantitative differences that mark continuous differences in task performance. The method of using qualitative differences in behavior to identify instances of conscious and unconscious perception was offered as an alternative to *a priori* attempts to establish a measure for consciousness (for original discussion see Reingold and Merikle 1988, 1990; see also Merikle and Daneman 1998 for review; Reingold 2004).

The ways that qualitative differences in behaviour have been related to subjective and objective measures of consciousness are discussed below in order to evaluate

---

This chapter contains parts of 'Signal detection theory, the exclusion failure paradigm and weak consciousness—Evidence for the access/phenomenal distinction?' from Consciousness and Cognition, Vol. 18, E. Irvine, pp. 551–560, Copyright 2009. Reprinted with permission from Elsevier.

DOI 10.1007/978-94-007-5173-6_3, © Springer Science+Business Media Dordrecht 2013

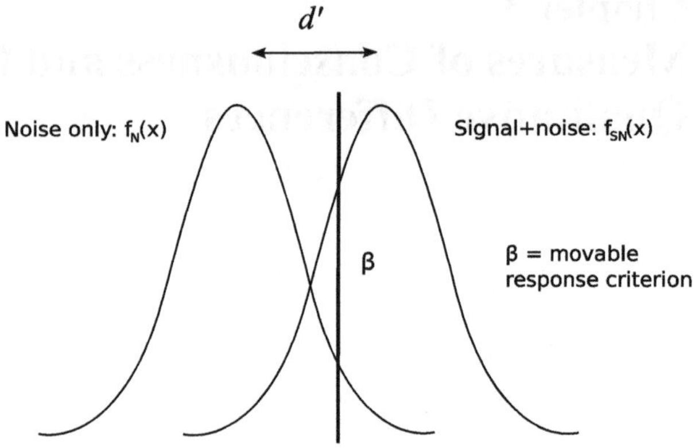

Fig. 3.1 Density functions of strength of internal responses to noise, $f_N(x)$, and signal + noise, $f_{SN}(x)$. Internal responses over β are reported as 'target present', responses below β are reported as 'target absent' by subjects

how useful this method can be in establishing a measure of consciousness. Through this, the problems in the way that the method of qualitative differences is used in consciousness science can be outlined. These problems concern how qualitative differences in behaviour are described, and how their relevance to a measure of consciousness is evaluated. In this chapter it is suggested that not only is it a serious methodological question of which measure of consciousness to use, but that the standard ways of moving forward on this question are simply not available in consciousness science. The suggestion that there are deep methodological problems in establishing *any* acceptable measure of consciousness is developed further in the next chapter.

3.2 Sensitivity d′ as a Measure of Consciousness

First, a brief reminder about the principles behind Signal Detection Theory and the objective measure *d′* it describes is necessary. The measure *d′* can be variously described as a measure of 'intrinsic discrimination acuity' or 'inherent accuracy' of a system to differentiate target signals of a particular amplitude or strength from a constant level of background noise. A system's sensitivity to a stimulus, *d′*, is given by the difference between the means of the density functions that determine how likely the system will make a 'target present' response in noise only trials ($f_N(x)$), and in signal plus noise trials ($f_{SN}(x)$), illustrated in Fig. 3.1. When the functions are the same due to internal processing constraints, the system cannot distinguish signals from noise, and $d' = 0$.

However, the measure *d′* does not by itself determine the responses that a system makes. Responses are modelled as decisions, constrained by a system's sensitivity, in which a system decides if the information available warrants a particular response in a given context. An internal criterion level β is "set intelligibly in accordance with the observer's perception of the prior probabilities of the two possible stimuli and of the various benefits and costs of correct and incorrect responses" (Swets 1996,

p. vii). The criterion level can change according to the task context and what kind of responses the system is trying to optimise.

In the example of perceptual defence mentioned in Chap. 1, subjects require longer stimulus durations to freely identify swear words than non-swear words. This is because they associate false alarms (falsely identifying a non-swear word as a swear word) with a cost, but will likely want to maximise the number of correct identifications of non-swear words. In this case subjects exhibit the same sensitivity to both swear and non-swear words, but different response criteria. In Fig. 3.1, the response criterion for swear words would be more conservative, shifting β to the right, while the response criterion for non-swear words would be more liberal, shifting it to the left.

In order to experimentally investigate the properties of unconscious perception, sensitivity *d*′ is often driven to zero by using short stimulus durations and masking, thus guaranteeing the elimination of conscious perception. This method was suggested as a way of combating the problems in early research into unconscious perception, caused by the use of subjective measures that allowed for the presence of some low level conscious perception (Blackwell 1952; Eriksen 1960; Goldiamond 1958; see also Holender 1986 for later discussions). Using subjective measures to find the threshold between conscious and unconscious perception does not guarantee that subjects are unconscious of stimuli, only that they do not report them.

Only *d*′ offers a way of eradicating response bias, and only when $d' = 0$ can it be safely ruled out that subjects are conscious of a stimulus. As a result, many view the setting of experimental parameters such that $d' = 0$ as a valuable method in unconscious perception research. Kouider and Dehaene's (2007) review of visual masking and unconscious semantic processing shows how careful use of *d*′ is now a standard part of experimental procedure in this area.

However, the use of *d*′ to guarantee the *absence* of consciousness does not mean that non-zero values of *d*′ are typically used to indicate the *presence* of consciousness (though see Snodgrass et al. 2004). To illustrate this, one of the standard cases used to argue against the use of *d*′ as a measure of (the presence of) consciousness is that of blindsight (Weiskrantz 1986). Patients with blindsight exhibit high *d*′ values, but do not report being conscious of stimuli and have low confidence in their responses in visual discrimination tasks. Despite the problems of relying on subjective reports, it is widely accepted that blindsighters have no (Type 1) or very little (Type 2) consciousness of stimuli in their 'blind' field (Azzopardi and Cowey 1998; Weiskrantz 1998). This is often taken as evidence that *d*′ is a good measure of a subject's sensitivity to stimuli, but not of conscious perception (see e.g. Lau 2008). Blindsight therefore appears to provide a clear dissociation between task performance as measured by *d*′, and consciousness.

However, this separation of task performance and awareness may not generalise to normal vision. The damage to V1 that causes blindsight has a number of effects on visual processing that are not present in normal visual systems. Normally, subjects exhibit the same sensitivity levels for yes/no and two alternative forced choice tasks (2AFC). These tasks are designed to measure the same level of availability of information, but responses are elicited in different ways. In yes/no tasks, subjects indicate whether or not a target is present, and in 2AFC tasks they have to choose between two responses that refer to a stimulus, e.g. whether the stimulus was presented in the top or bottom of a screen. Azzopardi and Cowey (1997, 1998) found that, in contrast to normally sighted subjects, blindsighter GY exhibited *different* sensitivities for yes/no tasks and two alternative forced choice (2AFC) tasks. Azzopardi and Cowey concluded that: "GY's residual vision is therefore different from normal vision near threshold, implying that his brain processes information about the visual stimulus in an unusual way" (1998, p. 308). The fact that d' clearly does not index consciousness in blindsighters does not necessarily mean that it fails to index consciousness in normals as well, as visual processing is clearly different in the two cases.

Blindsighters provide interesting cases in which reports are apparently unconnected with levels of d', and these cases are certainly very suggestive about what sort of measure to use. However, more general arguments against the use of d' as a measure of consciousness fail as they inherently rely on the assumption that subjective measures are instead the most appropriate measures of consciousness. This kind of argument is a question begging one: objective measures are clearly inadequate measures of consciousness, but only if it is already accepted that subjective measures are better. Unless there is independent reason to think that subjective measures are more appropriate (and there is rarely anything other than intuitions on offer), there is little in the way of a methodologically sound argument against objective measures of consciousness. Two arguments are assessed below to see if they can provide any methodological grounds for *favouring* objective measures of consciousness.

3.2.1 *d′, Decision-Making and Confidence Ratings*

Two lines of reasoning can be offered in favour of using objective measures of consciousness. These are based on identifying qualitative differences in behaviour that appear to map onto the distinction between conscious and unconscious perception. According to SDT analysis, differences in responses can be described as a product of both the subject's sensitivity to the stimulus and their response criterion. The contents of reports can be altered by changing the subject's response criterion, for example by altering their motivation. While differences in behaviour identified using subjective report-based measures are highly variable and easily manipulated, the differences in sensitivity identified by d' are highly stable and invariant. This suggests that the differences in behaviour captured by subjective measures (e.g. report vs. no report of a stimulus) are not the most relevant qualitative differences in task performance after all, as they are largely dependent on context and task demands. Given assumptions about the irrelevance of the placement of response criteria to a measure of consciousness, and the need for a stable measure, it seems that d' is a more appropriate measure of consciousness.

Further, there is the fact that subjects are capable (when pressed) of giving appropriate confidence ratings under all $d' > 0$ conditions (i.e. such that confidence correlates with accuracy, see Galvin et al. 2003), suggesting that d' indexes more than just information processing. It can also be argued that d' itself is based on responses about 'subjective' confidence. Responses to stimuli are based on how confident a system is that its internal representation of a stimulus is sufficiently strong to warrant a 'target present' response. The responses used to determine d' can therefore be seen as first order confidence ratings, compared with second order or Type 2 confidence ratings discussed in the previous chapter. If confidence is related to consciousness, then d' again seems like an adequate way of indexing consciousness, as $d' = 0$ identifies the threshold at which subjects can no longer make accurate Type 1 confidence ratings.

However, the qualitative differences appealed to as positive reasons to use d' to measure consciousness can be questioned. First, d' can only be seen as a better measure of consciousness if it is assumed that the factors that affect response generation (evidenced in subjective reports) are irrelevant to measuring consciousness. However, it is not clear if this assumption can be upheld. Instead, by seeing the whole research tradition that sprung out of the application of SDT to human perceptual systems as a research tradition about information processing only, it is easy to argue that objective measures are simply irrelevant to consciousness. This is more or less what the opponents of objective measures do (for history and review see Merikle at al. 2001). They argue that even though d' may be a stable and bias free measure of *something*, it is not clearly a measure of consciousness.

Instead, the decision-making evidenced in subjective reports is often taken as an essential part of consciousness, and therefore as an essential indicator of the presence or absence of consciousness. Dehaene et al. (2006) state that: "Conscious perception must... be evaluated by subjective report, preferably on a trial-by-trial basis" (p. 206). Lau's (2008) Higher Order Bayesian (HOB) decision-making theory of consciousness also explicitly states that "perceptual consciousness depends on the setting and maintaining of criteria based on representations of the statistical behaviour of internal signals" (p. 46). Clearly, researchers vary in how they think of the relationship between subjective reports and consciousness. Some view reportability as an experimental confound, and some view it as an essential marker of consciousness. So, it cannot simply be asserted that the bias free status of d' makes it a better measure than subjective measures.

Further, the stability of d' is also only a positive reason for using this measure if there are independent reasons for thinking that consciousness is a stable phenomenon itself, unaffected by expectation, motivation and context. Yet such independent reasons are not offered. Assertions about the adequacy of objective measures of consciousness rest on the assumption that response bias should be eliminated from measures of consciousness, and the assumption that the threshold between conscious and unconscious perception is a stable one. These assumptions, along with the converse assumptions that underlie the use of subjective measures are both unjustified, and cannot themselves be appealed to as support for either type of measure of consciousness.

It has also been claimed that subjects' ability to generate appropriate confidence ratings down to $d' = 0$ levels provides support for the claim that d' really measures consciousness rather than simple information processing. However, this raises the further problem of whether the generation of confidence ratings is relevant to a measure of consciousness. While more plausibly linked with the abilities of conscious subjects than 'mere' discrimination abilities, there is no reason why confidence ratings, as a measure of the evidence used to make decisions, could not be produced unconsciously. Kunimoto et al. (2001) who used SDT analysis of Type 2 confidence ratings to generate a measure of consciousness (discussed in the previous chapter) comment on this: "It could be argued that confidence judgments might be made without awareness just as discrimination judgments can be made without awareness... This could occur if subjects made their confidence judgments using specific strategies not based on awareness" (p. 304).

Indeed, this is the basic problem in consciousness science: using qualitative differences in behaviour in order to identify and measure consciousness is an untenable method unless it is clear that the behaviour in question is linked to consciousness. While some qualitative differences in behaviours may provide more intuitive measures than others, intuition in itself does not provide a methodologically sound reason for using these measures.

In this case, the basic question remains of how the method of searching for qualitative differences in behaviour can be used to identify a measure of consciousness. It is suggested below that the problem of identifying *relevant* qualitative differences always collapses into the problem of whether to use subjective or objective measures of consciousness. In this case, the method of qualitative differences may provide no evidence in support of a measure of consciousness independently of *a priori* assumption about the adequacy of either subjective or objective approaches. This suggestion is illustrated below through a discussion of measures of consciousness based on controllability.

3.3 Process Dissociation and Consciousness as Control

As a more advanced way of using qualitative differences in behaviour to establish a measure of consciousness, many researchers have followed Jacoby's (1991) process dissociation framework, used to distinguish automatic from intentional processes. Jacoby's work focuses on distinguishing implicit (automatic) and explicit (intentional) memory, but is intended to overcome the same methodological problems in distinguishing between unconscious (automatic) and conscious (intentional) perception. Typically, a task is devised to test the range of a particular ability associated with a type of memory or perception. Abilities that are linked to explicit memory or conscious perception are based on the notion of intentional control, such as the ability to overcome priming effects or engage in some other strategic type of task performance. The experimenter then looks for a qualitative difference in performance at this task. This difference is used to illustrate under what conditions a particular type of memory or perception is found.

For example, if explicit memory is associated with the ability to verbalise stored information, while implicit memory is associated with indirect priming effects only,

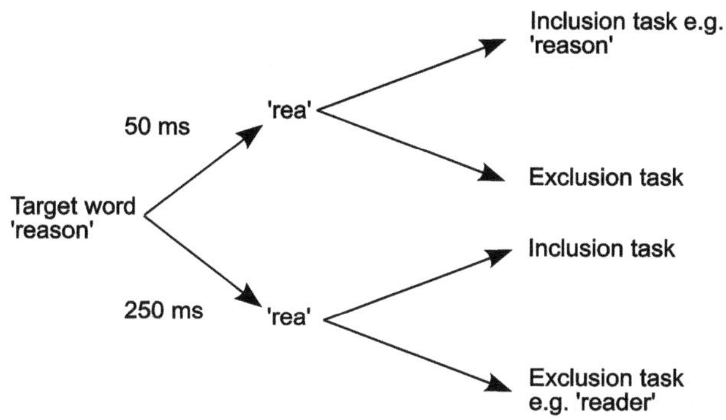

Fig. 3.2 The exclusion failure paradigm, with target word presented for either 50 or 250 ms. The inclusion task is to complete the word stem to form the target word (follow the primed response). The exclusion task is to complete the word stem to form a word that is *not* the target word (inhibit the primed response). In one variation of this paradigm, half the subjects are highly motivated by offered monetary reward for good performance

then a dissociation (a qualitative difference) in the ability to perform a word recall task will indicate the conditions under which explicit memory operates. Similarly, the conditions under which subjects are unable to intentionally control their responses to visual stimuli (such as primes) provides a dissociation in behaviour that could be used to mark the distinction between conscious and unconscious perception. Within Jacoby's framework, and that of researchers in unconscious perception (e.g. Merikle et al. 2001; Snodgrass et al. 2004), control or controllability is used as a way of differentiating conscious from unconscious perception. The equation of control with consciousness is analysed below to see if it can offer an alternative way of establishing a measure of consciousness.

3.3.1 Control as Inhibition

Given the basic idea that conscious perception enables some degree of control over behaviour, and that unconscious perception does not, there are a range of ways of operationalising 'control'. For example, Visser and Merikle (1999) identified the ability of subjects to successfully 'exclude' or inhibit primed responses in a word-stem completion task as a behaviour that could be used to identify conscious perception (see Fig. 3.2). Subjects are shown a target word (e.g. 'reason') for either 50 or 250 ms. The 250 ms duration is long enough for subjects to freely report the stimulus, but at 50 ms duration subjects do not report seeing the stimulus. A word stem 'rea' is then shown, and the subjects asked to perform either an inclusion or an exclusion task. The inclusion task is to complete the word stem to form the target word, 'reason'. This is an easy task to complete, even at 50 ms, as it is a case of standard priming. The exclusion task is to complete the word stem to form a word that is *not* the target—to exclude the target, to form for example 'reader'. This task is fairly hard even at 250 ms, and at 50 ms it becomes almost impossible.

Visser and Merikle argue that if a subject can inhibit primed responses when completing word stems, then they must be conscious of the prime stimulus, as control requires consciousness. A qualitative difference in subjects' ability to inhibit responses to the prime stimulus can therefore be used to assess the presence or absence of consciousness. As Visser and Merikle find such a dissociation between inhibitory and primed responses between the 50 and 250 ms prime duration, they claim that their method and measures can be used to identify instances of conscious and unconscious perception.

However, this paradigm makes use of a subjective, report-based measure of consciousness. As seen in the general criticisms of subjective measures of consciousness, free responses are open to response bias. In fact, with some money as motivation for good performance (Snodgrass 2002, 2004; Visser and Merikle 1999), or by making the task easier by providing two alternative responses to choose from (Fisk and Haase 2006), subjects become remarkably good at inhibiting primed responses under conditions in which they previously failed. The task and measure that Visser and Merikle use can therefore be criticized as they do not capture the whole range of conditions under which the ability associated with consciousness (inhibition of primed responses) can be exhibited. In this case, a dissociation in inhibitory capacities using subjective measures cannot be used to make inferences about the presence or absence of conscious perception.

Dissociations in controllability based on the bias-free measure d' cannot be criticized in this way however. This is because d' is constructed to index the full range of conditions under which an ability can be exhibited. The threshold at which subjects are no longer able perform the task, using objective measures of performance, is arguably the more relevant threshold for distinguishing conscious from unconscious perception, and supports the use of d' as a measure of consciousness. However, questions can also be raised about the validity of d' as a measure of consciousness, as noted above.

This points to a general problem with the process dissociation method. Dissociations in a particular type of behaviour can always be found at both the subjective (reportability) threshold and the objective ($d' = 0$) threshold. For example, for every behaviour, there is a threshold at which subjects cannot freely perform a task, and there is a different threshold at which subjects fail at the task when it is posed as a forced-choice task. Which threshold to use to distinguish conscious from unconscious perception depends on a prior commitment to subjective or objective measures of consciousness. Thus, while process dissociation methods were proposed as a promising new way of conducting consciousness research, they in fact collapse into an older and problematic debate about the relationship between subjective reports and sensitivity to a measure of consciousness.

One way of dealing with this problem is to reject the idea that we only want one measure of consciousness, and instead see these dissociations and measures as distinguishing different types of consciousness. As a neat way out of this methodological problem, this possibility is discussed below.

3.4 Qualitative Differences and Taxonomies of Consciousness

It is possible see the dissociations and measures discussed above as providing a taxonomic framework for different kinds of consciousness. For example, Block (2005) and Snodgrass (Snodgrass and Shevrin 2006; Snodgrass et al. 2004) claim that a Signal Detection Theory (SDT) approach to perception can be used to support Block's distinction between access and phenomenal consciousness. After a brief description of access and phenomenal consciousness, how Snodgrass and Block have used SDT to differentiate access and phenomenal consciousness will be explained. It is then argued that this attempt at using measures of consciousness to support a taxonomy of consciousness fails for the same methodological reasons described above. A discussion of just how serious these methodological problems are follows in the next section.

3.4.1 Access and Phenomenal Consciousness

As originally conceived by Block (1990), access consciousness concerns 'consumer' systems such as "systems of memory, perceptual categorization, reasoning, planning, evaluation of alternatives, decision-making, voluntary direction of attention, and more generally, rational control of action" (Block 2005, p. 48). In contrast, the 'what-it-is-like-ness' of consciousness is what Block calls phenomenal consciousness, or more recently, phenomenology (Block 2007). Block claims that we are conscious of more information than we can report or identify at any one time: "phenomenal consciousness overflows cognitive accessibility" (Block 2007, p. 481). The notion of phenomenal consciousness is difficult to operationalise because it cannot be identified via reportability, but Block and Snodgrass claim to be able to identify it using behavioural markers identifiable using SDT, as applied to the exclusion failure paradigm in particular.

SDT in itself is neutral about the relationship between subjective and objective measures and consciousness (Green and Swets 1966). However, it potentially provides a useful framework to use in a scientific description of conscious and unconscious perception, and qualitative differences in behaviour identified in SDT suggest themselves as markers for different kinds of consciousness. Thus Snodgrass and Shevrin (2006) have claimed that: "... objective threshold methods index phenomenally unconscious perception, whereas subjective threshold methods index phenomenally conscious but reflectively unconscious perception" (p. 74). They argue that above $d' = 0$ but below the subjective threshold (i.e. not currently reported), subjects are phenomenally conscious of stimuli but do not have cognitive access to this information. With a shift in the subject's response criterion, this information can become accessed and reported. Although Block does not support Snodgrass's model of conscious/unconscious perception in full, he does assert that in some cases, such as Visser and Merikle's exclusion failure paradigm outlined above, SDT analysis suggests that there are cases of phenomenal consciousness without cognitive access.

3.4.2 Exclusion Failure, Phenomenal and Access Consciousness

In addition to Visser and Merikle's (1999) main exclusion failure paradigm described above (see Sect. 3.3.1), they also investigated the effects of motivation on performance by offering monetary rewards for good task performance to half of their subjects. They found that those who were highly motivated to perform well at the task did significantly better at excluding (inhibiting) prime words for the 50 ms duration than those who were not. According to Visser and Merikle this effect was due to increased attention making previously unconscious information consciously available, and therefore allowing conscious inhibitory control.

In contrast, Block (2005) and Snodgrass and Shevrin (2006; see also Snodgrass et al. 2004) argue that reportability is not an adequate measure of (phenomenal) consciousness. Instead, they claim that increased motivation makes stimuli more reportable by shifting the subjects' report criterion, but that the stimuli were consciously perceived both with and without motivation. That is, in normal conditions subjects are phenomenally conscious of the 50 ms stimuli but do not access this information to inform their responses. With increased motivation, subjects use a less conservative response criterion and make use of (access) their phenomenally conscious perceptions to exclude the target word.

Both Block and Snodgrass claim that the exclusion failure paradigm provides support for the distinction between unreported phenomenal consciousness and reported access consciousness: "There is, therefore, evidence in the 'exclusion' case of experiential contents (e.g. as of seeing 'reason') without the kind of access required for report, planning, decision-making, evaluation of alternatives, memory and voluntary control of attention" (Block 2005, p. 49). According to Snodgrass, subjects' criterion levels for reportability can be used as a measure of access consciousness, and their sensitivity to d', which indexes potential reportability, can be used to measure the presence or absence of phenomenal consciousness.

The problem with both of these interpretations of the exclusion failure paradigm, and the taxonomy of consciousness that Block and Snodgrass suggest, is that there is a simpler alternative explanation available, based on SDT itself. Fisk and Haase (2006) have argued that failures to inhibit primed responses for the 50 ms stimulus duration are simply indicative of subjects failing to complete a difficult task. They found that if the exclusion task is made easier for subjects by asking for a two-alternative forced choice decision (choice between two possible words), instead of a free report, subjects are able to perform above chance on the exclusion instruction around a 50 ms duration. Using a similar masked semantic priming task, Bengson and Hutchison (2007) found that exclusion success varied as a product of response criteria, manipulated by changing task instructions. They note that "... exclusion failure... may reflect a participant's *decision* not to exclude briefly flashed information, rather than *inability*" (p. 787).

In this case the original finding of exclusion failure using subjective measures does not stem from subjects being unconscious of the 50 ms stimuli, (Visser and Merikle), or that perceptual information of the 50 ms stimuli is just not available in access consciousness (Block and Snodgrass). Exclusion failure (measured by free report) is determined by the subject's placement of their response criteria and their internal decision-making process in a particular context. None of this reflects the availability of information in access consciousness, only how it is used. Failures to exclude a primed response are consistent with subjects being 'access conscious' of the prime, but simply not using this information (e.g. due to the placement of their response criterion). Whether information is actually used to inform a response, and whether it is available to be used to inform a response, are not the same thing. The case of perceptual defence discussed earlier provides a clearer case of this; both swear and non-swear words are present in access consciousness, but only one set of words is consistently used to inform subjects' responses. In this case, reportability (subjective measures) and sensitivity (objective measures) cannot be seen as direct analogues to access and phenomenal consciousness (for further discussion see Irvine 2009).

This highlights the general need for consciousness researchers to be more explicit about which phenomena and measures they refer to. In the case of subjective and objective measures of consciousness, this is decision-making and report, and sensitivity. Yet in this case it may simply be easier to label any resulting taxonomy of measures as a taxonomy of alternative approaches in psychophysics, rather than as a taxonomy of consciousness. Reasons why this route may in fact be the only methodologically sound one are developed below, and extended in the next chapter.

3.5 Which Qualitative Differences? Differences in What?

> From qualitatively different experimental effects one generally infers distinct underlying processes. Assuming that qualitatively different effects are indeed observed... this fact does not in itself constitute evidence for [the conscious/unconscious distinction]... because qualitatively different effects can be observed in the processing of consciously identifiable stimuli. (Holender 1986, p. 3)

The method of identifying qualitative differences in behaviour in order to provide an adequate measure (or measures) of consciousness has been traced through several different ways of operationalising consciousness. One is that $d' = 0$, the threshold of sensitivity, identifies the most appropriate difference in behaviour as it indexes the threshold above which successful stimulus detection can occur. It also provides the threshold above which confidence ratings and inhibitory and strategic responses can be generated. Alternatively, the subjective threshold, assessed through reportability, may identify the most appropriate qualitative difference in behaviour for a measure of consciousness, as it seems to map onto our intuitions about consciousness in a more direct way. Discussions of how relevant either of these thresholds are to consciousness illustrates an underlying methodological problem.

The first part of the problem is that the qualitative differences discussed above are typically not transparently described. The ways that subjective measures of task performance can be manipulated illustrate that they are most relevant to identifying the variables that affect decision-making and criterion setting. In contrast, differences in the ability to perform a task above chance or not, established through SDT, may simply assess the information processing capacity of system. Differences in controllability invite investigations into the mechanisms that underlie the inhibition of automatic responses. Providing *accurate descriptions* of these qualitative differences in behaviour is essential if they are to be associated with a methodologically sound measure of consciousness. However, described in this way, none provide a qualitative difference that is obviously related to the distinction between conscious and unconscious perception.

This is largely because the proper description of these behavioural differences makes it difficult to identify a *relevant* difference by which to measure consciousness. By analysing the ability that a task assesses (e.g. inhibition), and the way it is assessed in the task (e.g. through subjective or objective approaches), it is possible to question whether a qualitative difference in behaviour really is between conscious and unconscious perception, or between two other processes. For example, if a qualitative difference can be described as a product of 'merely' shifting response criteria, or 'mere' information processing, then its ability to distinguish conscious from unconscious perception appears weaker.

The second part of the problem is that for whatever behavior we identify, there will be a qualitative difference in behaviour found at both subjective *and* objective thresholds. The only justification that can be given for using one measure rather than another is by stating untestable pre-theoretical assumptions about the properties of consciousness. Those who view response criteria as a confounding factor in measures of consciousness will use objective measures. Those who view response criteria and report as an essential part of consciousness will use subjective measures. These are two very different conceptions of where along a chain of processing consciousness occurs, and thus how it should be measured. So, empirical work can provide evidence that a qualitative difference in behaviour is a significant difference in *something*, but it is not conclusive about whether this qualitative difference can be used to measure consciousness.

This also affects attempts to use a range of qualitative differences in behaviour to generate taxonomies of consciousness. Block and Snodgrass et al. use the exclusion failure paradigm to identify phenomenal consciousness with failures of inhibition below the response criterion, and access consciousness with successful acts of inhibition above the response criterion. However, this attempt to outline a taxonomy of conscious states fails due to an alternative interpretation of the difference, based on SDT itself. Instead of failures of inhibition being evidence for a lack of access to conscious information, they instead show how the placement of response criteria affects what information is used to make a decision. The qualitative difference between successful and unsuccessful inhibition is not therefore a relevant one to indexing types of consciousness based on cognitive access.

However, there are clearly other debates about how to measure or define a phenomenon in science, and perhaps ways of resolving these debates could be helpful here. For example, scientific entities and processes are typically defined in terms of the current most popular means of operationalizing them, and measured by appealing to convergence towards particular markers. So, given a standard way of producing a phenomenon, it can be measured using a marker (or set of markers) that gives the most reliable (i.e. replicable) values. However, as argued above, both sides in this debate (over subjective and objective measures) have essentially the same amount of evidence in their favour. For every given ability associated with consciousness, there are two possible markers (subjective and objective measures), and two dissociations (at subjective and objective thresholds). There is convergence for both markers (though objective measures may be more stable over subjects), and routine use of both in experimental situations. Neither a standard theoretical framework or experimental paradigm, nor a single convergent marker, can be appealed to.

Another possible resolution is for researchers to come to a pragmatic agreement on which behaviours and measures to use to investigate consciousness. However, given the 'special' nature of consciousness, this seems unlikely; there is serious debate, based on conflicting intuitions and pre-theoretical commitments, about what behaviours can be taken as markers of consciousness, and thus how consciousness should be measured. Consensus is not a feature of consciousness science, and neither is it seen as a viable way to solve methodological problems.

Therefore, the problem is *not* simply that disagreement exists about how to measure consciousness, but that the ways of progressing from this disagreement do not seem to be open in consciousness science. There are pre-theoretical reasons, and evidence of qualitative differences in behaviour, that support the use of both subjective and objective measures of consciousness equally well. Developing methodological worries that stem from such a stalemate, and evaluating their implications on consciousness science, is continued in the following chapters.

3.6 Conclusions

The examples discussed above and in the previous chapter show that there are significant methodological problems in identifying and measuring consciousness. Measures of consciousness are typically based on qualitative differences in behaviour, but there is a methodological impasse in the debate over whether to use subjective or objective measures of consciousness. The aim of establishing a taxonomy of consciousness based on these measures also appears untenable.

Chapter 4 expands on the problems identified in this chapter, including a discussion of the essential heuristic role that the method of qualitative differences typically plays in science. The method of using qualitative differences in observable properties to identify and demarcate phenomena is a usually a productive one, and forces constant refinement of descriptions and taxonomies of phenomena. However, it will be argued that this does not happen in consciousness science, and that this is why there are such deep and ongoing debates about how to establish a measure of consciousness. These arguments contribute to the final conclusion that 'consciousness' cannot function as a viable scientific concept, as the application of scientific methods in consciousness science do not comply with the basic standards of scientific practice.

References

Azzopardi, P., & Cowey, A. (1997). Is blindsight like normal, near-threshold vision? *Proceedings of the National Academy of Sciences USA, 94*, 14190–14194.

Azzopardi, P., & Cowey, A. (1998). Blindsight and visual awareness. *Consciousness and Cognition, 7*, 292–311.

Bengson, J. J., & Hutchison, K. A. (2007). Variability in response criteria affects estimates of conscious identification and unconscious semantic priming. *Consciousness and Cognition, 16*, 785–796.

Blackwell, H. R. (1952). Studies of psychophysical methods for measuring visual thresholds. *Journal of the Optical Society of America, 42*, 606–616.

Block, N. (1990). Consciousness and accessibility. *Behavioral and Brain Sciences, 13*, 596–598.

Block, N. (2005). Two neural correlates of consciousness. *Trends in Cognitive Science, 9*, 46–52.

Block, N. (2007). Consciousness, accessibility, and the mesh between psychology and neuroscience. *Behavioral and Brain Sciences, 30*, 481–548.

Dehaene, S., Changeux, J. P., Naccache, L., Sackur, J., & Sergent, C. (2006). Conscious, preconscious, and subliminal processing: A testable taxonomy. *Trends in Cognitive Sciences, 100*, 204–211.

Eriksen, C. W. (1960). Discrimination and learning without awareness: A methodological survey and evaluation. *Psychological Review, 67*, 279–300.

Fisk, G. D., & Haase, S. J. (2006). Exclusion failure does not demonstrate unconscious perception II: Evidence from a forced-choice exclusion task. *Vision Research, 46*, 4244–4251.

Galvin, S. J., Podd, J. V., Drga, V., & Whitmore, J. (2003). Type 2 tasks in the theory of signal detectability: Discrimination between correct and incorrect decisions. *Psychonomic Bulletin and Review, 10*, 843–876.

Goldiamond, I. (1958). Indicators of perception: 1. Subliminal perception, subception, unconscious perception: An analysis in terms of psychophysical indicator methodology. *Psychological Bulletin, 55*, 373–411.

Green, D. M., & Swets, J. A. (1966). *Signal detection theory and psychophysics*. New York: Wiley.

Holender, D. (1986). Semantic activation without conscious identification in dichotic listening, parafoveal vision, and visual masking: A survey and appraisal. *Behavioral and Brain Sciences, 9*, 1–23.

Irvine, E. (2009). Signal detection theory, the exclusion failure paradigm and weak consciousness—Evidence for the access/phenomenal distinction? *Consciousness and Cognition, 18*, 551–560.

Jacoby, L. L. (1991). A process dissociation framework: Separating automatic from intentional uses of memory. *Journal of Memory and Language, 30*, 513–541.

Kouider, S., & Dehaene, S. (2007). Levels of processing during non-conscious perception: A critical review of visual masking. *Philosophical transactions of the Royal Society B, 362*, 857–875.

Kunimoto, C., Miller, J., & Pashler, H. (2001). Confidence and accuracy of near-threshold discrimination responses. *Consciousness and Cognition, 10*, 294–340.

Lau, H. C. (2008). A higher order Bayesian decision theory of consciousness. In R. Banerjee & B. K Chakrabati (Eds.), *Progress in Brain Research, 168*, 35–48.

Merikle, P. M., & Daneman, M. (1998). Psychological investigations of unconscious perception. *Journal of Consciousness Studies, 5*, 5–18.

Merikle, P. M., Smilek, D., & Eastwood, J. D. (2001). Perception without awareness: Perspectives from cognitive psychology. *Cognition, 79*, 115–134.

Reingold, E. M. (2004). Unconscious perception and the classic dissociation paradigm: A new angle? *Perception & Psychophysics, 66*, 882–887.

Reingold, E. M., & Merikle, P. M. (1988). Using direct and indirect measure to study perception without awareness. *Perception and Psychophysics, 44*, 563–575.

Reingold, E. M., & Merikle, P. M. (1990). On the inter-relatedness of theory and measurement in the study of unconscious processes. *Mind and Language, 5*, 9–28.

Snodgrass, J. M. (2002). Disambiguating conscious and unconscious influences: Do exclusion paradigms demonstrate unconscious perception? *The American Journal of Psychology, 115*, 545–579.

Snodgrass, J. M. (2004). The dissociation paradigm and its discontents: How can unconscious perception or memory be inferred? *Consciousness and Cognition, 13*, 107–116.

Snodgrass, J. M., & Shevrin, H. (2006). Unconscious inhibition and facilitation at the objective detection threshold: Replicable and qualitatively different unconscious perceptual effects. *Cognition, 101*, 43–79.

Snodgrass, J. M., Bernat, E., & Shevrin, H. (2004). Unconscious perception: A model-based approach to method and evidence. *Perception and Psychophysics, 66*, 846–867.

Swets, J. A. (1996). *Signal detection theory and ROC analysis in psychology and diagnostics: Collected papers*. Mahwah: Erlbaum Associates.

Visser, T. A. W., & Merikle, P. M. (1999). Conscious and unconscious processes: The effects of motivation. *Consciousness and Cognition, 8*, 94–113.

Weiskrantz, L. (1986). *Blindsight: A case study and implications*. Oxford: Oxford University Press.

Weiskrantz, L. (1998). Consciousness and commentaries. *International Journal of Psychology, 33*, 227–233.

Chapter 4
Dissociations and Consciousness

4.1 Introduction

As discussed in the previous chapter, qualitative differences have been used to investigate and measure consciousness. In particular, it was discussed whether the qualitative differences in behaviour observed at the objective (sensitivity) or the subjective (report-based) threshold could be used to index and measure consciousness. This chapter is concerned with more fundamental problems in how dissociations in behaviour, which include qualitative differences, are interpreted in consciousness science. This constitutes a shift away from examining specific examples towards the methodological question of whether dissociations provide *any* evidence that can be used to establish measures of, or definitions for, consciousness.

Dissociations are seen as an important, though problematic, way of investigating the structure and function of systems, often used in cognitive psychology (Shallice 1988; Vallar 1999). A single dissociation between performance levels at two different tasks occurs when an experimental manipulation affects performance on one task but not the other. A simple example of this is putting a blindfold over someone's eyes, and finding that while their ability to read is diminished, their ability to discriminate sounds is not. This provides evidence that the systems required for reading and auditory discrimination are in some way independent. A much stronger dissociation is a double dissociation. In this case, two different experimental manipulations each affect performance at two different tasks in contrasting ways. For example, blindfolding someone inhibits their ability to read but not to discriminate sounds, while wearing earplugs inhibits their ability to discriminate sounds but not to read. In this case, there is evidence that (some part of the) systems required for reading and auditory discrimination are functionally independent. However, only single dissociations will be addressed in this chapter as it seems fundamentally unlikely that a double dissociation could be found between conscious and unconscious perception (as it would require the existence of 'higher level' conscious perception without 'lower level' unconscious processing).

Dissociation methodology is used in consciousness science as a way of identifying and investigating the range of conscious and unconscious perception. If a dissociation can be found between a putative measure of conscious perception and a measure

E. Irvine, *Consciousness as a Scientific Concept,* Studies in Brain and Mind 5,
DOI 10.1007/978-94-007-5173-6_4, © Springer Science+Business Media Dordrecht 2013

of unconscious perception, then the conditions under which the dissociation occurs can tell us something about the properties of consciousness. Also, the measures used to establish the dissociation might be reasonable ways of measuring conscious and unconscious perception outside the dissociation paradigm. Yet the history of unconscious perception research, as well as more general theoretical work on dissociation methods, shows that there are many problems associated with establishing and interpreting dissociations (Reingold and Merikle 1988, 1990; Schmidt and Vorberg 2006 for review and analysis).

These problems also prompt an analysis of the heuristic role of dissociations in cognitive science, compared with consciousness science. This will be explored through a comparison with the controversial interpretation of the function and independence of the ventral and dorsal streams of visual processing (e.g. Milner and Goodale 1995, 2008; Schenk and McIntosh 2010). The comparison highlights the differences in the ways that dissociations are typically established, how they are interpreted, and the ways in which they are used to guide research, in standard scientific contexts compared with consciousness science. In particular, the comparison highlights a failing to use the details of experimental manipulations to clearly identify the dissociated phenomena, and to test and progressively revise how the dissociation is interpreted. It is therefore suggested that using the categories of conscious and unconscious perception to interpret empirical evidence is methodologically empty, and that these categories should not play a role in scientific theorising.

4.2 Single Dissociations and Measures of Consciousness

In consciousness research dissociations are sought between direct measures of perception, D, used to assess conscious perception, and indirect measures of perception, I, used to assess unconscious perception. Direct measures assess the ability of a subject to make a particular kind of response towards a target stimulus, for example to detect it. Indirect measures I measure the effect of the perception of one stimulus on the ability to respond to a different target stimulus. These 'indirect' abilities are typically tested using priming tasks, in which the effects of perceiving an 'unconscious' prime can be measured through a subject's response towards a subsequent target. Cases in which indirect measures outstrip direct measures, $I > D$, constitute a dissociation between direct and indirect measures of perception. In standard cases, dissociations are sought between an objective measure of detection, using d' as a direct measure D of consciousness, and a measure of priming as an indirect measure I of unconscious perception. When there is evidence of priming without evidence of detection, then it is claimed that the priming is the result of unconscious perception. This evidence is also used to identify some of the properties of conscious and unconscious perception and, importantly, a way of measuring them.

However, merely finding a dissociation between D and I does not ensure that they are adequate measures of conscious and unconscious perception, as D and I need not satisfy any stringent criteria in order to be used to establish a dissociation.

In fact, they could be measures of both unconscious and conscious perception to a greater or lesser degree. That is, measures could partially index both conscious and unconscious perception, but in a way that allows them to exhibit dissociations in particular circumstances. In this case, (rather surprisingly), the dissociated measures that are used to illustrate the distinction between conscious and unconscious perception cannot necessarily be used as measures of conscious and unconscious perception (Erdelyi 1986; Schmidt and Vorberg 2006).

This state of affairs reflects two competing aims of consciousness science. One is to establish the properties of both conscious and unconscious perception. The second is to provide an adequate measure of consciousness, as explored in the previous chapters. While the two appear to be intimately linked, dissociations may not be able to fulfill either aim. The relationship between different types of single dissociations and measures of consciousness is explored below.

4.2.1 Single Dissociations and Measures of Consciousness

In order to establish a single dissociation between direct (*D*, conscious) and indirect (*I*, unconscious) measures of perception, two reasonably weak constraints must be met by the measures *D* and *I*. One is that the measures must be weakly monotonic measures of the phenomena they are used to index. This means that the measures must at least stay the same, or increase, with an increase in the ability or phenomenon measured (e.g. detection ability or priming effects). The other assumption is that the direct measure *D* is at least as sensitive to conscious information as the indirect measure *I*. This means that the direct measure must track conscious perception better than the indirect measure.

A typical example of a direct measure *D* and an indirect measure *I* that satisfy these constraints is referred to above; *D* is often a measure of a subject's ability to detect a target, while *I* is often a measure of priming effects. *D* can be safely assumed to be a weakly monotonic measure of conscious perception as detection rates increase with stimulus duration (and therefore plausibly with conscious perception of the stimulus), and *D* is also plausibly more sensitive to conscious perception than a measure of priming. Given these constraints, if a dissociation is found such that $I > D$ then *I* must have been influenced by some information other than conscious information (see Fig. 4.1). This is taken to be evidence for the existence of unconscious perception (Reingold and Merikle 1988).

Establishing a single dissociation between a direct measure D and an indirect measure *I* does not however show that *D* is an otherwise usable measure of consciousness. According to the constraints above, both *I* and *D* can measure aspects of both conscious and unconscious perception. In this case, while a single dissociation can provide evidence of the existence of two different perceptual processes, it does not necessarily provide a usable measure of consciousness. Indeed, Reingold (2004), commenting on earlier work on a relative sensitivity paradigm used to establish just

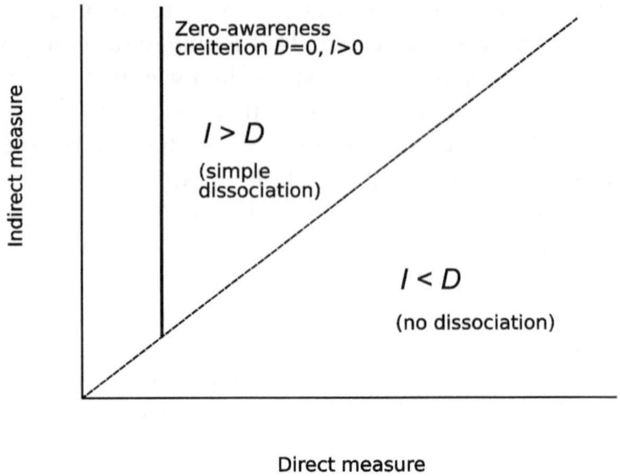

Fig. 4.1 Single dissociations (adapted from Fig. 4.1, p. 493, Schmidt and Vorberg 2006). The zero-awareness criterion is a special case ($D = 0$) of a simple dissociation, for which $I > D$. For the criteria that each must satisfy, see text. Reused with permission from Springer

the kind of single dissociation noted here, states that he and Merikle "... were careful to note that the RSP cannot resolve the fundamental quest for a valid measure of consciousness or awareness" (Reingold 2004, p. 883).

However, the zero-awareness criterion offers a different way of establishing single dissociations and measures of consciousness. The zero-awareness criterion is used to identify a subset of single dissociations for which $I > D$, where the direct measure D is assumed to be an adequate measure of consciousness, and is set to zero ($D = 0$). Then all that is required to establish a dissociation is for the indirect measure I to be greater than the value of D, that is, $I > 0$ (see Fig. 4.1). This kind of single dissociation is used by researchers to identify properties of conscious perception, based on the measure D and the conditions under which $D = 0$, and to identify instances of unconscious perception.

As seen in the previous chapter, and from the example above, this method is far more common. The standard practice of testing for unconscious perception while $d' = 0$ is an example of a zero-awareness dissociation, and qualitative differences are just zero-awareness dissociations. For example, the exclusion failure paradigm uses the ability to inhibit primed responses as a direct measure D, and the exhibition of priming as the indirect measure I. In cases where subjects do not inhibit primed responses ($D = 0$), they still exhibit primed responses ($I > 0$). As subjective measures of response inhibition are assumed to adequately measure conscious perception, this dissociation is described as a dissociation between conscious and unconscious perception. However, in order to establish that a zero-awareness criterion dissociation does actually identify a distinction between conscious and unconscious perception, and not two other types of perception, an additional constraint on D must first be met.

As stated by Reingold and Merikle (1988, 1990; also Reingold 2004), for D to be an adequate measure of conscious perception, and for zero-awareness dissociations to be dissociations between conscious and unconscious perception, D must be an *exhaustive* measure of consciousness. This means that D measures all aspects and all instances of conscious perception. When such a measure is set to zero, the only thing that could be measured by non-zero values of the indirect measure *I* would be some other form of perception (i.e. unconscious perception). If *D* fails in these criteria, then it is possible that when $D=0$, there is still the possibility that conscious perception is present. In this case, non-zero values of the indirect measure *I* could simply reflect instances of conscious perception, and thus not provide a dissociation between conscious and unconscious perception.

Accordingly, most of the methodological work in unconscious perception research has focused on providing an exhaustive direct measure *D* of consciousness, and of ensuring that it is set to zero in zero-awareness paradigms. As suggested in the previous chapter, while report-based measures exhaustively measure reportability (by definition), they do not obviously exhaustively measure the information subjects are conscious of. What subjects report is affected by response bias, so subjects may only report the presence or identity of stimuli that they are reasonably confident about. For this reason, objective measures such as SDT sensitivity d' are now often used to find dissociations between conscious and unconscious perception as they are seen as more exhaustive measures of conscious perception. Setting $d'=0$ is itself a difficult practical problem, and has received much attention in experimental work (e.g. Holender 1986; Kouider and Dehaene 2007 for review).

However, while the objective measure d' may be an exhaustive measure, dissociations found using d' do not necessarily cleanly dissociate conscious and unconscious perception. This is because objective measures such as d' may not *exclusively* index conscious perception, so may measure many instances of unconscious perception too. In this case, perception that occurs at $d'=0$ is unconscious perception, but perception that occurs above $d'=0$ is not necessarily conscious perception. This is well known in the literature, but underlines the general point that dissociations may not be as informative as they are often thought to be.

In order to provide a direct measure that can not only be used to establish a dissociation between conscious and unconscious perception, but that also allows inferences to be made about the properties of both conscious and unconscious perception, the direct measure must be an exhaustive *and* an exclusive measure of consciousness. That is, a direct measure must index all, and only, instances of conscious perception. The problems with satisfying these constraints, as suggested in the previous chapters, are significant. There are persuasive arguments that subjective measures are exclusive but not exhaustive measures of consciousness (due to response bias some instances of consciousness may not be captured by report). There are also persuasive arguments that objective measures are exhaustive but not exclusive measures of consciousness (they also potentially measure instances of unconscious perception).

Yet in consciousness science it is often (implicitly) assumed that the dissociation methodology can be used to establish a measure of consciousness. The underlying idea is that if only we could find a dissociation between two measures of task performance that was sufficiently intuitive or persuasive enough, then we could use those measures to index conscious and unconscious perception. This was seen in the previous chapter where qualitative differences and dissociations were sought between reportability, sensitivity, control, and different types of confidence ratings as a way of anchoring a measure of consciousness. However, the plain existence of these dissociations does little to settle the question of what measures of consciousness we should use.

No dissociation in itself can validate a measure as an adequate (i.e. exhaustive and exclusive) measure of consciousness; these are criteria that must be justified independently. Even comparative claims that one measure is more exhaustive than another may not be true as claims about exhaustive measures of *consciousness*. For example, sensitivity d' is a more exhaustive measure than reportability *of something*, given that the objective threshold $d' = 0$ is often significantly lower than reportability thresholds. However, the dissociations that can be found using these measures do not themselves show what it is, that is more exhaustively measured by d', and less exhaustively measured by reportability. What this something is requires interpretation. The general constraints on how dissociations can be interpreted, and the implications for the categories of conscious and unconscious perception, are discussed below.

4.3 Interpreting Dissociations

> Dissociations seem to imply a partition, but a partition of what? (Dunn and Kirsner 2003, p. 4)

Dissociations are used to infer that there are at least two components or processes that contribute to the dissociated measures. However, one of the major obstacles in using dissociations is to establish what exactly can be inferred from the dissociation. A dissociation shows the existence of two different *somethings*, but these somethings cannot be identified by looking to the dissociation alone. Instead, an interpretive framework is needed to conceptualise what these somethings are. For example, direct and indirect measures of perception are associated with conscious and unconscious perception, but this is merely an assumption, or part of a model.

Importantly, dissociations themselves cannot be used to support one interpretive framework over another. As Dunn and Kirsner (2003) state: "... while a dissociation may be interpreted as signifying something within an *a priori* conceptual framework, they do not constitute, in themselves, evidence for this framework" (p. 4). Schmidt and Vorberg (2006) also comment on this idea in relation to consciousness science, and state that dissociations in themselves are relatively uninformative: "... any two dissociated sources of information are conceivable, each of which may be conscious or unconscious (or maybe of yet another type)" (p. 501).

The problem is that dissociations tell you very little about how they should be interpreted. Dissociations between direct and indirect measures of perception can be interpreted as dissociations between conscious and unconscious perception, or they can be interpreted more simply as dissociations between the processes that underlie forced choice detection performance and priming effects (or perhaps as something else entirely). Whether these interpretations come to the same thing is an open question.

Typically, the range of possible interpretations of dissociation evidence means that these interpretive frameworks are constantly questioned, tested, revised, or other alternatives suggested. That is, what the dissociated phenomena *are*, is not predetermined or set in stone. Yet in consciousness science the interpretive framework of conscious vs. unconscious perception is assumed to be the right one through which to interpret dissociations between direct and indirect measures of perception. This may seem like an unsurprisingly and relatively unproblematic result of the basic assumptions behind consciousness science, but it will be argued below that questioning interpretive frameworks is a crucial research heuristic that is lost in consciousness science.

4.3.1 Dissociations and Interpretive Frameworks: Ventral and Dorsal Perceptual Streams

To illustrate the way in which dissociations are typically used in a heuristic role it is easiest to consider an example. One particularly interesting and controversial claim, based on dissociations in behaviour as a result of localised lesions, is that the dorsal stream engages in 'vision-for-action' and that the ventral stream engages in 'vision-for-perception' (Goodale and Milner 2004; Milner and Goodale 1995; Ungerleider and Mishkin 1982). Early dissociation evidence in monkeys based on object discrimination and spatial proximity tasks suggested that the dorsal stream was used in spatial tasks, while the ventral stream was used for object identification: "It has been our working hypothesis that the ventral or occipitotemporal pathway is specialised for object perception (identifying *what* an object is) whereas the dorsal or occipitoparietal pathway is specialised for spatial perception (locating *where* an object is)." (Ungerleider and Mishkin 1982, p. 549). The range of tasks that monkeys could and could not do with different lesions, described in terms of object vs. spatial discrimination tasks, contributed directly to the characterisation of the function of the processing streams that were dissociated.

However, different dissociation paradigms were later used by Milner and Goodale (1995) to question the characterisation of the function of the two streams. In human subjects the tasks that are inhibited by lesions in dorsal and ventral streams were found to be more specific than suggested by earlier experiments. Lesions in the dorsal stream appear to affect spatial judgement as it is used to guide actions, but subjects are able to make accurate spatial judgments independently of actions. Lesions in the ventral stream appear to affect higher cognitive processes such as object identification, but subjects are able to interact normally with objects in their environment.

Instead of dissociations being evidence for a distinction between spatial and object-based perception, Milner and Goodale describe the function of the ventral and dorsal streams in terms of the different roles that they play in action selection and implementation. In a recent paper (2008), they describe the function of the ventral stream as identifying appropriate goal objects in the environment, represented relative to other objects in the environment (allocentric coding). In contrast, they characterise the function of the dorsal stream as implementing actions to achieve a goal provided by the ventral stream, with the necessary visual information represented in an egocentric format.[1]

Thus, a number of new paradigms have been used to test behavioural dissociations in perceptual abilities. These have been used to update and redefine the framework that is used to interpret the dissociation evidence, and therefore how the functions of the dorsal and ventral streams are characterised. Shifting from a simple distinction between object-based and spatial perception, dissociations are now used to support distinctions in perceptual processing which both rely, but in different ways, on object and spatial information. The details of the tasks used to establish these dissociations are crucial in describing the dissociated phenomena, and therefore in forming the frameworks through which these dissociations are interpreted. Further dissociations can be tested to provide evidence for this framework, or for alternative frameworks (e.g. Dijkerman et al. 2009; McIntosh et al. 2011).

This iterative process of interpreting dissociations, devising experimental paradigms to test interpretive frameworks, and then revising the framework in light of new evidence, provides a crucial heuristic for advancing research. Importantly, a core feature of this heuristic is that the dissociation paradigms and the frameworks used to interpret them are subject to constant change.

4.3.2 Dissociation Methods in Consciousness Science: The Exclusion Failure Paradigm

The research heuristic associated with the use of dissociation methods is ignored in consciousness science. Unlike the way in which the interpretive frameworks described above have been empirically tested, revised, and tested again, leading to radical changes in how the dissociated phenomena are categorised, researchers in consciousness science do something very different. They assume that conscious and unconscious perception are the 'right' categories to use to interpret dissociation evidence, and try to fit experimental evidence around these categories. Crucially, the practice of referring to precise experimental manipulations and tasks in order to inform an interpretive framework does not occur in consciousness science. The basic interpretive framework goes untested and unrevised. The exclusion failure paradigm is used to illustrate this below, followed by a discussion of the interpretation of proposed measures of consciousness in general.

As discussed earlier, the exclusion failure paradigm has been used to provide a zero-awareness dissociation between direct measure D of subjects' ability to inhibit primed responses, and the indirect measure I of priming effects, in a word-stem

[1] Links between the ventral/dorsal distinction and consciousness are ignored in this discussion, as they naturally provide problems of their own, but see Milner (1995) for original suggestion and Jacob and de Vignemont (2010) for recent work.

completion task. This dissociation has been interpreted as a dissociation between conscious and unconscious perception (i.e. $D = 0$, $I > 0$). However, as discussed in Chap. 3, applications of Signal Detection Theory show that it is simply a dissociation between reportability and priming effects, based around the placement of subjects' response criteria.

This interpretation of the exclusion failure paradigm does not make reference to conscious or unconscious perception because they are not categories that inform the structure of the task, its variants, or the measures used. The factors that affect D, where D is a measure of reportability, are factors that affect response bias, as tested using the kind of experimental manipulations noted above. Dissociations using report based measures are therefore best interpreted in terms of reportability. Similarly, the manipulations used to achieve $d' = 0$ make no reference to consciousness, only to task performance at forced-choice detection. It is only forced-choice detection that should therefore figure in the interpretation of dissociations using d'. It is a standard feature of dissociation paradigms that interpretive frameworks refer to the tasks and measures used.

Clearly though, interpretive frameworks often go further than just a description of the experimental evidence. In this case, one could be warranted in interpreting dissociations in task performance as dissociations between consciousness and unconscious perception, even if this does not directly reflect the tasks and measures used to establish the dissociation. However, in going beyond a fine-grained description of the experimental manipulations and measures used, interpretive frameworks do not usually go far, and are always testable. Crucially, the predictions these frameworks generate refer to further specific experimental manipulations and measures. For example, the functions attributed to the ventral and dorsal streams are directly based on the fine-grained details of the dissociated phenomena, but also describe general patterns found in experimental data, such as the apparent need for the egocentric coding of spatial relations in the dorsal stream. Experimental paradigms can then be devised to test this interpretive framework, with these experimental findings feeding back into the statement of the framework.

However, the interpretation of dissociations between direct and indirect measures of perception as dissociations between conscious and unconscious perception does not refer to the experimental manipulations and measures used. Neither is this interpretation directly testable using further experimental manipulations. Testing whether a dissociation does in fact identify the difference between conscious and unconscious perception involves manipulating the original task, or coming up with new tasks. However, these further tests are only informative, relative to the conscious/unconscious distinction, if we have already assumed that (for example) controllability is a relevant task, and subjective measures of task performance are adequate measures of consciousness. Without these assumptions, task manipulations and further tests are completely uninformative with regards to the distinction between conscious and unconscious distinction. In contrast, no such pre-theoretical commitments can be found in the dorsal/ventral case discussed above. Further, as seen in earlier chapters, these commitments are difficult to justify on theoretical grounds.

This means that within dissociation methodology, no real work is done by the labels of conscious and unconscious perception. Whatever predictions or further research questions that are generated using these labels are parasitic on the tasks used to operationalize consciousness, measures of reportability and sensitivity, and our view of their relation to consciousness. Tagging on a label of 'conscious' or 'unconscious' perception to the real distinctions that are uncovered using these methods is not a valid use of dissociation methodology.

4.4 Tasks, Measures, Manipulations and Interpretive Frameworks

The examples above illustrate how the standard method of continually testing and refining interpretations of dissociated phenomena is missing from consciousness science. Instead of the products of research being used to provide a more fine-grained description of dissociated phenomena, empirical results are fitted around pre-theoretical intuitions about what the phenomena under investigation are, and how they should be measured (e.g. via report or d'). Yet in general the intuitions that foster the use of categories of mental phenomena are not sufficient in themselves to validate them as scientifically useful distinctions.

For example, following a discussion of memory research Bechtel (2008) states: "When a variety of mental operations have been convincingly identified and localised in the appropriate brain areas, it may turn out that the characterisation of the operations are orthogonal to these long-standing categories of mental phenomena" (Bechtel 2008, p. 82). Likewise, it is necessary to take the products of consciousness research for what they really are, and analyse them to see whether the categories of conscious and unconscious perception are appropriate ones to use to characterise dissociated phenomena. The examples above suggest that the categories of conscious and unconscious perception are unlikely to capture scientifically useful distinctions, as compared with categories that refer explicitly to the experimental manipulations and measures that generate the dissociations.

In general, a failure to refine interpretive frameworks in light of experimental work is a significant methodological problem. Researchers do not typically use coarse-grained and ill-defined categories to interpret dissociations for long. Perception research has used dissociations to provide fine-grained categories of motion, spatial, colour, object and scene perception, all further sub-divided, along with conditions that affect each of these types of perception. It no longer makes sense to ask whether a subject sees a stimulus; the question is far too vague. Only by making reference to the specific type of perception and a specific measure can a question be posed. By finding dissociations between an ability to perceive in one way (e.g. forced choice detection) but not another (e.g. forced choice identification), research becomes more fine-grained and better able to describe the specific conditions required for the exhibition of a specific ability. Over time, research questions that refer to older, broader categories of phenomena (such as 'seeing an object'), no longer function as well-posed questions, or at least have a range of answers. The fact that researchers can still ask whether a subject is conscious of a visual stimulus, without referring to any of the well-known, fine-grained distinctions of perception, is at odds with how dissociation methodology typically plays out.

However, some researchers do try to adapt their theories and taxonomies of consciousness to the dissociations they uncover, for example by distinguishing between categories like phenomenal and access consciousness (Block 2005; Lamme 2006; Snodgrass and Shevrin 2006; Snodgrass et al. 2004), or between first order and self-consciousness. But the point can be rallied against these apparently more empirically informed categories too. In Chap. 3 it was argued that the exclusion failure cannot be used to distinguish between phenomenal and access consciousness, only between different ways of placing response criteria and making decisions. For other 'types' of consciousness there will also be a way of describing them in the language of cognitive science, that makes reference to the precise experimental manipulations used to establish dissociations.

The continual imposition of the categories of conscious and unconscious perception, and continual debates over how to measure consciousness, result only from our determination that these are viable scientific categories of phenomena. This leads to a special problem: the standard ways of building on disagreements in order to identify, describe, and measure phenomena do not seem to apply in consciousness science. What qualitative difference and dissociation paradigms illustrate is that there are a range of things that subjects sometimes can, and sometimes cannot do. Dissociations provide evidence about the limits on a subject's ability to do these specific things (e.g. identify words, exclude primed words) according to different measures. These dissociations can be iteratively interpreted by referring to experimental manipulations and results, but at no point do these dissociations make contact with 'consciousness', only a range of different sensory and cognitive phenomena that different people associate with consciousness. Dissociations themselves can tell us nothing about the relation between these different sensory and cognitive phenomena and consciousness.

Using dissociation evidence to argue about the distinction between conscious and unconscious perception, or between different types of consciousness, is to force an artificial debate onto ill-fitting evidence. Instead, dissociations and qualitative differences allow inferences to be made about the structure of phenomena like perceptual processing and decision-making, in terms that appeal directly to the details and results of experimental paradigms. Ignoring the heuristic roles that dissociation methods play in cognitive science not only generates an artificial debate about how to distinguish and how to measure consciousness, but it also blinds researchers to what we can actually learn from dissociation methods.

4.5 Conclusions

The aim of this chapter has been to challenge the assumptions that dissociation methods can provide adequate measures of conscious and unconscious perception, and that these are useful and viable categories with which to interpret dissociations in perception research. Instead, dissociations themselves provide no guide as to how to satisfy the exhaustivity and exclusivity constraints on finding an adequate measure of consciousness. Further, it is only by looking in detail at the experimental manipulations used to establish a dissociation that viable interpretations of dissociation phenomena can be formed. As seen in Chap. 3, it is necessary to provide clear descriptions of qualitative differences in behaviour, including a description of the task and measure used in a particular paradigm, in order to consider them as evidence for particular claims about consciousness. However, these descriptions show that far from providing support for a particular characterisation or measure of consciousness, qualitative differences provide evidence only for claims made about reportability or sensitivity, or different types of sensory of cognitive processing.

Further, this chapter has showed that using the categories of conscious and unconscious perception severely limits the important heuristic role that dissociation methodology typically exhibits in scientific research. This is to use experimental evidence to continually inform the frameworks through which a dissociation is interpreted. Dissociations themselves provide no interpretive framework, but these frameworks can be generated by considering the tasks and measures used to establish the dissociation. These frameworks can then be tested and continually revised by further, often more fine-grained, experimental work. Labelling dissociations as dissociations between conscious and unconscious perception goes hard against this practice, as it imposes an interpretive framework that is not warranted by experimental work, and that is not directly testable. By ignoring the heuristic value of dissociation methods, consciousness science starts to look distinctly unscientific.

The rest of the book builds on these ideas, examining research that uses concepts of consciousness, and testing whether the use of these concepts is consistent with scientific methods. Aside from dissociation-based methods, it is shown how other methods are misused or ignored in order to preserve the categories of conscious and unconscious perception. The next chapter revisits the goal of establishing a measure of consciousness through the alternative strategy of engaging in multi-level integrative research. However, arguments will be offered to show that integrative methods in consciousness science fail to fulfill two essential preconditions for successful integration and convergence; that of having independent evidence, and that of operationalising suitably similar phenomena. Similar to the claims made in this chapter, it will be argued that the only convergence that can be found across proposed measures of consciousness stems from the similarity of the sensory and cognitive phenomena that are actually measured. It will be further suggested that there are no reasons why these measures should be described as convergent measures of *consciousness*.

References

Bechtel, W. (2008). *Mental mechanisms: Philosophical perspectives on cognitive neuroscience*. London: Routledge.

Block, N. (2005). Two neural correlates of consciousness. *Trends in Cognitive Science, 9*, 46–52.

Dijkerman, H. C., McIntosh, R. D., Schindlre, I., Nijboer, T. C. W., & Milner, A. D. (2009). Choosing between alternative wrist postures: Action planning needs perception. *Neuropsychologia, 47,* 1476–1482.

Dunn, J. C., & Kirsner, K. (2003). What can we infer from double dissociations? *Cortex, 39,* 1–7.

Erdelyi, M. H. (1986). Experimental indeterminacies in the dissociation paradigms of subliminal perception. *Behavioral and Brain Sciences, 9,* 30–31.

Goodale, M. A., & Milner, A. D. (2004). *Sight unseen: An exploration of conscious and unconscious vision.* Oxford: Oxford University Press.

Holender, D. (1986). Semantic activation without conscious identification in dichotic listening, parafoveal vision, and visual masking: A survey and appraisal. *Behavioral and Brain Sciences, 9,* 1–23.

Jacob, P., & De Vignemont, F. (2010). Spatial coordinates and phenomenology in the two-visual systems model. In N. Gangopadhyay (Ed.), *Perception, action, and consciousness: Sensorimotor dynamics and two-visual systems* (pp. 125–144). Oxford: Oxford University Press.

Kouider, S., & Dehaene, S. (2007). Levels of processing during non-conscious perception: A critical review of visual masking. *Philosophical transactions of the Royal Society B, 362,* 857–875.

Lamme, V. A. F. (2006). Towards a true neural stance on consciousness. *Trends in Cognitive Sciences, 10,* 494–501.

McIntosh, R. D., Mulroe, A., Blangero, A., Pisella, L., & Rosetti, Y. (2011). Correlated deficits of perception and action in optic ataxia. *Neuropsychologia, 49,* 131–137.

Milner, A. D. (1995). Cerebral correlates of visual awareness. *Neuropsychologia, 33,* 1117–1130.

Milner, A. D., & Goodale, M. A. (1995). *The visual brain in action.* Oxford: Oxford University Press.

Milner, A. D., & Goodale, M. A. (2008). Two visual systems re-viewed. *Neuropsychologia, 46,* 774–785.

Reingold, E. M. (2004). Unconscious perception and the classic dissociation paradigm: A new angle? *Perception & Psychophysics, 66,* 882–887.

Reingold, E. M., & Merikle, P. M. (1988). Using direct and indirect measure to study perception without awareness. *Perception and Psychophysics, 44,* 563–575.

Reingold, E. M., & Merikle, P. M. (1990). On the inter-relatedness of theory and measurement in the study of unconscious processes. *Mind and Language, 5,* 9–28.

Schenk, T., & McIntosh, R. D. (2010). Do we have independent visual streams for perception and action? *Cognitive Neuroscience, 1,* 52–63.

Schmidt, T., & Vorberg, D. (2006). Criteria for unconscious cognition: Three types of dissociation. *Perception & Psychophysics, 68,* 489–504.

Shallice, T. (1988). *From neuropsychology to mental structure.* Cambridge: Cambridge University Press.

Snodgrass, J. M., & Shevrin, H. (2006). Unconscious inhibition and facilitation at the objective detection threshold: Replicable and qualitatively different unconscious perceptual effects. *Cognition, 101,* 43–79.

Snodgrass, J. M., Bernat, E., & Shevrin, H. (2004). Unconscious perception: A model-based approach to method and evidence. *Perception and Psychophysics, 66,* 846–867.

Ungerleider, L. G., & Mishkin, M. (1982). Two cortical visual systems. In D. J. Ingle, M. A. Goodale, & R. J. W. Mansfield (Eds.), *Analysis of visual behavior* (pp. 549–586). Cambridge: MIT Press.

Vallar, G. (1999). The methodological foundations of neuropsychology. In G. Denes & L. Pizzamiglio (Eds.), *Handbook of clinical and experimental neuropsychology* (pp. 95–131). Hove: Psychology Press.

Chapter 5
Converging on Consciousness?

5.1 Introduction

As seen in previous chapters, there is currently a wide array of measures and theories of consciousness on offer. Recently, neurophysiological markers have been proposed as alternative measures of consciousness, using new techniques to assess the time course, strength and location of brain activity. These markers include different kinds of neural synchrony, early and late ERPs, and local and global recurrent processing. Somewhat predictably, there is as much disagreement over which neurophysiological marker should be used as a measure of consciousness as there is over which behavioural marker to use. Further, as discussed in earlier chapters, dissociation methods on their own seem unable to identify an appropriate measure of consciousness.

However, there are different approaches available. One is to make the most of the wide range of behavioural and neurophysiological measures available and compare them with each other to establish their advantages and disadvantages. In Seth et al.'s (2008) review of measures of consciousness, it is suggested that "... an integrative approach combining both types of measures in a single study encourages a virtuous circularity in which putative measures and theoretical advances mutually inform, validate and refine one another" (p. 320). Seth et al. argue that by using multiple measures in the same experimental paradigms, the strengths and weakness, and similarities or differences between them, can be found.

By comparing multiple measures, it should be possible to establish which measures best fit intuitive, practical and theoretical constraints on a measure of consciousness. These constraints may concern how sensitive, reliable, or consistent a measure is, how broad its potential use is, or how well it fits with our expectations about what results a measure should give in certain situations. The comparative strengths of different measures can be built on and refined, leading to better measures. Shea and Bayne (2010) have also proposed that a similar method, that of searching for convergent groups of markers, should be used to identify appropriate measures of consciousness.

The proposal that Seth et al. make of using inter-level refinement to converge on appropriate measures and theories of consciousness is echoed in recent work in

E. Irvine, *Consciousness as a Scientific Concept*, Studies in Brain and Mind 5,
DOI 10.1007/978-94-007-5173-6_5,

philosophy of neuroscience. In particular it can be found in reductive (Bickle 2006) and multi-level (Bechtel 2008; Craver 2007) accounts of explanation in neuroscience. Both types of account rely on comparing and integrating research carried out over multiple levels of description, in order to converge on a coherent explanation of a phenomenon. The comparison of behavioural and neurophysiological markers of consciousness both across levels of description (e.g. between behavioural and neurophysiological markers), but also within the same level of description (e.g. between neurophysiological markers), clearly constitutes an example of this integrative method.

Importantly, all integrative approaches rest on the same two assumptions. One is that the range of experimental methods used to operationalise a phenomenon, either within or across multiple levels of description, really do operationalise the same phenomenon. They may operationalise different aspects of the phenomenon by probing it in different ways, but in order for convergence and coherence to emerge, the experimental methods used must all be ways of probing the same common phenomenon. Convergence among measures cannot emerge if fundamentally distinct phenomena are measured across different experimental paradigms. These accounts also rely on the assumption that evidence gained across multiple levels of research is independent, such that integrating this research can be informative.

This chapter addresses the viability of an integrative approach in consciousness science, and whether or not such an approach will lead to convergence among measures of consciousness. This will be done in three stages. First, the framework developed in Sullivan's (2009) assessment of the integrative approaches found in philosophy of neuroscience will be used to characterise the different ways that experimental paradigms operationalise consciousness. This framework is based on an analysis of experimental practice and offers a way of assessing whether different experimental paradigms do, in fact, operationalise the same phenomenon. The second stage will be to outline the kinds of convergence that can and cannot be found across behavioural and neurophysiological measures of consciousness. The lack of independence between many behavioural and neurophysiological measures of consciousness will be highlighted, thus showing the limitations of this approach. Where convergence can or cannot occur between independent measures will be shown to be predictable, given the way that consciousness is operationalised in particular paradigms. Finally, it will be suggested that there is little reason to view the convergent measures as measures of aspects or types of consciousness. Instead, they are measures of the various phenomena operationalised in specific experimental paradigms; a range of distinct sensory, cognitive, and neural phenomena that have little in common.

5.2 A Pre-Condition for Successful Integration and Convergence

First, it is essential to consider the constraints on successful integrative approaches. One such constraint is that in order to find any convergence within a set of integrated measures, there must be some common features, or some common phenomenon, that they all assess. If the phenomena they measure have little in common, then they are unlikely to behave in similar ways under a range of experimental manipulations. In this case, measures of these phenomena will not converge.

One way of investigating how likely it is that a set of experimental paradigms operationalise the same or similar phenomena, and thus how likely it is that convergent measures will emerge, is to consider the range of variation present across the paradigms. If experimental paradigms are sufficiently different, and different in ways that can be recognised as being significant, then there is room to question whether they really are probing the same phenomenon. Of course variation across experimental paradigms does not in itself rule out the possibility of convergence. Indeed, having multiple ways of probing the same phenomenon is usually of great value as it allows researchers to investigate the different properties of a phenomenon, and to test how it fares under a range of conditions.

However, it is standard practice that by looking at the differences across experimental paradigms and the results they produce, this information can be used to identify distinctions and differences between different phenomena (e.g. using dissociation methods). By identifying important differences across experimental paradigms, it is possible to argue that a range of different phenomena are operationalised under the same name, in which case there is little chance of an integrated, convergent model or explanation emerging. The range of variation, and across which variables, that serves to demarcate different phenomena will clearly depend on the specific phenomena under investigation, the specific experimental paradigms used, and background information from relevant research fields. The range of acceptable and unacceptable variation in experimental protocols relative to one phenomenon, semantic priming, can be used to illustrate this.

Different experimental paradigms are used to investigate the effects of masked primes, or primes shown for short durations, on semantic processing of subsequent targets. Accuracy rates and response times in word categorisation and word association tasks can be used to test how the processing of prime words affects how target words are processed. For example, in testing the semantic processing of numbers (Dehaene et al. 1998), a number prime can effect the accuracy and reaction times of subjects who must categorise a target number as being greater or smaller than 5. Primes that are in the same semantic category as the targets (congruent trials) give rise to more accurate performance and shorter reaction times, while primes in a different semantic category to the target result in lower accuracy and longer reaction times.

Responses in word-stem completion tasks can also be used to show the effects of primes on word generation. Following the example used in Chap. 3 and 4, primes are shown to subjects (e.g. ‘reader’), followed by a word stem (‘rea’), and subjects are required to complete the word stem to either form the prime word (inclusion instruction), or a different word (exclusion instruction). Subjects tend to be more successful and faster in completing the task if they are asked to form the prime word rather than a different one (Visser and Merikle 1999). Several types of measure (accuracy, response times) can be used in several different types of experimental paradigms (word categorisation, word-stem completion) to assess the conditions under which primes affect semantic processing.

There are several reasons why it is assumed that these different experimental paradigms really are operationalising the same phenomenon, and why they are viewed as a set of convergent results. One is that there is an obvious similarity between measures. Accuracy and response times are recognised as generally being highly correlated, (e.g. see use in Bentin et al. 1985; Holcomb 1993; Perea and Gotor 1997). Also, manipulations of prime duration or masking (within certain bounds), and the semantic category of the prime, all give rise to the same pattern of results across these paradigms. Finally, the paradigms have been explicitly developed to manipulate a very specific set of processes, based on background knowledge about neural processing. All the tasks above can be modeled via a common process of (something like) spreading activation (Collins and Loftus 1975). The processing of primes biases the processing of semantic information of subsequent target words towards words similar to the prime. Background knowledge is used to identify relevant variable that indicate where distinctions and similarities are likely to be found in semantic processing. In this case, the variation across the experimental protocols described above are not seen to compromise the assumption that they all operationalise the common phenomenon of semantic priming.

However, some variations in experimental paradigms mean that they clearly operationalise different phenomena. For example, if semantic priming were to be tested by subjects first having to respond to the prime, and then the target, or if the delay between the prime and that target was sufficiently long, these variations would be recognised as being significant enough that they would operationalise different phenomena. This is because attentional allocation is a variable that affects whether semantic processing of a target occurs at all due to the attentional blink (Raymond et al. 1992), and the variation of prime-target delay also affects whether semantic priming occurs. Due to the differences in these experimental paradigms, their failure to generate the same patterns of results as other semantic priming paradigms, and background knowledge of which other phenomena they are likely to operationalise, they would be recognised as testing different phenomena (e.g. attentional blink, semantic processing of non-primed target). Variation *per se* across experimental paradigms does not rule out convergence across measures. However, significant amounts of variation in relevant variables often means that different phenomena are operationalised and measured, thus preventing convergence.

Sullivan's (2009) assessment of the likelihood of convergence and successful integration in contemporary neuroscience provides a more detailed framework with which to investigate variation across experimental paradigms (also referred to as experimental protocols), which will be used in the following sections. This includes an analysis of production procedures (details of stimulus presentation), measurement procedures (what variable is measured) and detection procedures (how the measures are interpreted in relation to the phenomenon under investigation).

Given the variations in production, measurement and detection procedures she identifies, Sullivan argues that researchers using different experimental protocols to investigate a particular neuroscientific phenomenon in fact often operationalise a range of (sometimes very different) phenomena. For example, in research into the molecular foundations of social recognition memory in mice, different behavioural tasks have been used to assess the presence or absence of social recognition memory following a molecular intervention. Due to the differences between these behavioural protocols, and background knowledge of how important these differences are likely to be, Sullivan argues that the molecular mechanisms that are investigated using these protocols are quite different.

Similarly, in research into long-term potentiation (LTP, related to memory), she identifies the different ways that LTP is stimulated, including a range of different strengths, durations, delays and numbers of electrical pulses to different parts of a neuron. Given the sensitivity of neuronal and synaptic activity to different types of electrical stimulations, it is unlikely that the range of production procedures used to generate LTP in fact stimulate the same mechanism or process. In this case there is no single phenomenon identified in LTP research. Occasionally the differences in experimental protocols are noted when different labs find different results, but often they are not. Instead, a range of labs that each use their own sets of experimental protocols all claim to provide different competing explanations of the 'same' phenomenon. She comments that "an investigation of the multitude of experimental protocols in the LTP field alone suggests that it has predominantly been and actually still remains an unconstrained free-for-all" (Sullivan 2009, p. 529).

By making explicit the differences between experimental protocols in these examples, Sullivan aims to show that the assumption that they do in fact operationalise the same phenomenon is made on weak grounds. In fact, once brought to light, these differences are clearly significant ones, and can be used to explain the range of contrasting results found in neuroscientific research. Based on the claim that these experimental protocols operationalise different phenomena, Sullivan argues that no unified or integrative explanations can be expected for these phenomena. This is because "it is not clear that neuroscientists working within the same field are even talking about the same phenomenon" (Sullivan 2009, p. 525).

Given the material discussed in the previous chapters, and the material discussed below, it should become increasingly clear that consciousness science is also an 'unconstrained free-for-all'. Sullivan's framework will be used to identify the differences in experimental protocols found in neurophysiological investigations of consciousness, from which suggestions will be made about where distinctions and similarities between protocols are likely (or are) to be found. Given this, it will be argued that the protocols used to operationalise consciousness in fact operationalise a range of very different phenomena, suggesting that (interesting) convergence across putative measures of consciousness is unlikely to emerge.

5.3 Neurophysiological Measures of Consciousness

Behavioural measures of consciousness have been discussed in earlier chapters, and below is an outline of some neurophysiological markers that have been proposed as measures of consciousness. This includes an exploration of the ways in which neurophysiological measures are dependent on behavioural measures, or on theoretical commitments about what sort of phenomenon consciousness is likely to be. The measures outlined below are not the only ones available, and are assessed as ways of investigating the contents of consciousness rather than 'levels' of consciousness (e.g. sleep vs. wakefulness). However, an assessment of these measures alone is sufficient to illustrate the problems in expecting convergence between a range of behavioural and neurophysiological measures of consciousness. The three neurophysiological measures considered here are Event Related Potentials (ERPs), neural synchrony and recurrent processing.

ERPs are waveforms of electrical activity that are measured using EEG and MEG across a wide area of the scalp (though more precise EEG measures can be taken using localisation information provided by fMRI). ERPs track neural activity over time with a high degree of temporal resolution, with different kinds of waveforms being found for different parts of cognitive tasks (e.g. attentional allocation, object identification, etc). In consciousness science, there is debate over which of the temporal sequence of ERP waveforms can be used as markers of consciousness, some favouring earlier ERPs, and some later.

Neural synchrony is a measure of how well, in what frequency, and for how long, neurons fire together in different parts of the brain. The existence of neural synchrony is essential for EEG techniques to register any useful markers, as large amounts of unco-ordinated activity would cancel each other out. By analysing the relative strength of measured activity across different frequency bands, researchers are able to identify instances of neurons firing together in different oscillatory patterns. Further evidence of neural synchrony localised to particular brain areas can also be derived from fMRI data by investigating the functional connectivity between local brain areas, done by analysing covariance in the BOLD signal (for reviews on neural synchrony, how it is measured, and current uses see e.g. Uhlhaas and Singer 2006; Uhlhaas et al. 2009). Strong and sustained neural synchrony in higher frequency (gamma) ranges for longer periods of time in frontal areas of the brain were originally claimed to be markers of conscious processing (Crick and Koch 1990). More recent research has emphasised the important role of earlier, transitory and lower frequency (alpha, beta) neural synchrony in 'conscious' task conditions.

Finally, recurrent processing refers to evidence of information being processed both in a forward and backward direction (i.e. recurrently), which is seen as evidence of flexible rather than automatic processing. The existence of recurrent processing in the brain can be inferred from a mix of anatomical investigations of cortical connectivity (e.g. see Lamme and Roelfsema 2000, for review), as well as EEG and fMRI. Using EEG to establish the time course of a particular set of processing, fMRI can be used to localise brain activity. Activity that continues to occur in early brain areas (found using fMRI) after information has been passed to later brain areas (as suggested by EEG data), supplemented by knowledge of cortical connections, suggests that information is still being transferred between the early and late brain areas via recurrent processing.

For example, Scholte et al. (2008) used EEG to track the time course of processing, as well as fMRI to locate where neural activity occurred, in order to establish whether V1 contributed to scene segmentation through recurrent processing loops. With a slightly different technique, Koivisto et al. (2011) used fMRI-guided transcranial magnetic stimulation (TMS) to 'knock out' selected brain areas during an object categorisation task. They showed that activity in early visual areas (V1/V2) is required *after* information has reached later brain areas (area LO) in order to successfully complete the task, suggesting that recurrent processing across these areas is necessary for object categorisation. While standard global workspace theories (e.g. Baars 1997; Dehaene et al. 2006) claim that recurrent processing across frontal-parietal as well as sensory areas is necessary for consciousness, some argue that recurrent processing in local sensory areas is sufficient for some types of consciousness (Lamme 2006).

Unlike the contrast between subjective and objective behavioural measures, there are no fundamental problems in integrating different neurophysiological measures. Indeed, there is a high degree of convergence between them, as they are all based on the same set of methods to isolate temporal and spatial markers of neural activity related to specific tasks. The reliance on neural synchonry in all the measures above ensures that they are all related, though synchrony may differ in its location, frequency, duration, strength, and time-course.

Instead, what is relevant to the possibility of an integrative approach in consciousness science are the debates *amongst* proponents of each type of measure. These debates are about whether local or global, or early or late, neurophysiological markers are the most appropriate ones to use as measures of consciousness. Thus there is a high degree of similarity between arguments for late ERPs, late neural synchrony, and global recurrent processing as the best measures of consciousness. Likewise, the contrasting arguments in support of early ERPs, early neural synchrony, and local recurrent processing are very similar. Despite the overlap in arguments for late/global and early/local activity across all three measures, each measure is considered separately below, as this is the way they are typically discussed in the literature, though similarities in early/late and local/global debates are highlighted throughout.

5.3.1 Early vs. Late ERPs

By looking at differences in ERP sequences for 'seen' and 'unseen' stimuli several research groups have argued that either early or late ERP waveforms can be used as measures of consciousness. Two main research groups have suggested that an early ERP signal correlates with consciousness. Koivisto and Revonsuo (2003) found that a negative ERP peaking at 200 ms differentiated conscious from unconscious perception in a change blindness paradigm (subjects have to do a 'spot the difference' task on two similar scenes serially presented). Koivisto et al. (2005) found that early ERPs for awareness occurred before attentional selection, and Koivisto et al. (2006) supported this by finding further evidence of an early negative ERP (130–320 ms after stimulus onset) that occurred independently of the scope of attention (local or global). Pins and Ffychte (2003) have argued that even earlier ERPs (100 ms after stimulus onset) are correlated with consciousness. Using fMRI and measuring evoked potentials, they found that early activity in occipital lobes correlated with consciousness in subjects detecting an unmasked grating.

In contrast, Dehaene's group have argued that while these early ERPs contribute to the later conscious states, they are not themselves markers of consciousness. They use evidence from attentional blink and backward masking paradigms, in which subjects must identify a letter presented after a specific time interval from a mask (said to close down or 'blink' attentional resources), or identify letters that are presented before a mask. They found that only later ERPs, primarily the P3 waveform, correlate with consciousness. Sergent et al. (2005) found that the N3, P3a and P3b ERPs (around 270–300 ms) correlated with consciousness in the attentional blink paradigm. Del Cul et al. (2007) and Dehaene et al. (2001) have also found evidence for the correlation of the P3 with consciousness in masking paradigms.

The underlying reasons for the debate between proponents of early and late ERPs can be found in the different production and detection procedures used, as well as the differences in the underlying theoretical assumptions made about consciousness across different research groups. First, some researchers identify the occurrence of consciousness with the earliest significant differences between ERPs for 'seen' and 'unseen' trials, in paradigms that generate a continuous report distribution. Using this detection procedure, early ERPs are identified as markers of consciousness. Based on the distinction between phenomenal and access consciousness, Koivisto and Revonsuo 2003 (Koivisto et al. 2006) also used this detection procedure to argue that early ERPs correlate with phenomenal consciousness, while later ERPs (such as the P3) correlate with access consciousness.

However, using the same detection procedure, but different conceptions about the nature of consciousness Pins and Ffychte (2003) come up with a rather different conclusion. They view the time-series of ERP signals as all contributing to a conscious experience: "... the relative timing of different nodes argues against a unitary process related to the perception of the grating. Instead, it suggests a segregation of function across the network with each node performing a different perceptual/cognitive operation" (Pins and Ffychte 2003, p. 472).

Dehaene and colleagues use a different detection procedure altogether. They use subjective visibility ratings as behavioural markers of consciousness, and have found that there seems to be an 'all or nothing' response to both backward-masked stimuli and for stimuli in the attentional blink. Instead of looking for *any* significant differences between ERP signals for 'seen' and 'unseen' trials, only ERPs that match the bimodal 'all or nothing' groups of the visibility ratings are classed as correlates of consciousness. Although divergence between earlier ERPs can be found for 'seen' and 'unseen' trials, only later ERPs, primarily the P3, clearly show this *bimodal* distribution. Therefore they suggest that based on this detection procedure, only later ERP components are markers of consciousness, while earlier ERPs only form necessary background conditions: "... while those early components may contribute to the subsequent transition toward conscious access or to its failure, they do not yet correspond to a full-blown conscious state" (Del Cul et al. 2007, p. 2420).

So, if consciousness is separated into phenomenal consciousness and access consciousness (Koivisto), then early and late ERPs provide a useful way of identifying the stages of processing that map to these two categories. Alternatively, if consciousness is seen as the sum total of a range of processes over time (Pins and Ffychte), then early and late ERPs mark different functional activities that contribute to the overall phenomenon of consciousness. In contrast with these views, Dehaene and colleagues argue that consciousness occurs at a later stage of processing that matches bimodal report distributions, in which case it can be identified when, and only when, later ERPs are found.

Different views on what sort of phenomenon consciousness is, and how this affects the use of different production and detection procedures, can therefore lead to different interpretations of early and late ERPs as measures of consciousness. In particular, the way in which neurophysiological markers are based on (sometimes controversial) behavioural markers suggests that neurophysiological markers of consciousness cannot escape the criticisms made about behavioural investigations of consciousness. That is, neurophysiological markers based on different types of responses (e.g. continuous vs. bimodal response distribution, objective vs. subjective responses) are liable to pick out correlates that reflect different kinds of processing. Unless it is accepted that a particular kind of behavioural response is a good marker of consciousness, the accompanying neurophysiological marker will also be a questionable marker of consciousness. This problematic relationship between behavioural and neurophysiological measures of consciousness is repeatedly found in the discussion below.

5.3.2 Early Transient vs. Late Sustained Neural Synchrony

There is also a debate about whether early and transient, or late and sustained, neural synchrony over particular frequency ranges is the best marker of consciousness. Several studies have found correlations between early and transient neural synchrony and 'seen' trials. However, it is less clear than in the ERP cases what conclusions to draw from this. For example, Melloni et al. (2007) found that early transient gamma-band synchrony correlated with 'seen' trials in a 2AFC delayed match-to-sample task but noted that "it remains to be clarified whether the early large-scale synchronisation is already the neuronal correlate of phenomenal awareness or whether awareness emerges only from the entirety of the processes following this coordinated state" (Melloni et al. 2007, p. 2864). Fries et al. (2002), and Palva et al. (2005), who found evidence of an important role for early alpha-band synchrony, as well as later synchronous activity in beta and gamma bands in 'seen' trials, make similar comments.

So, while some of these authors do not claim that early synchrony is a clear marker of consciousness, they suggest that it is at least necessary for the later stages of sustained synchrony. For example, Melloni et al. (2007) suggest that long distance, gamma-band synchronisation "plays a role in triggering the cognitive processes associated with conscious awareness" (p. 2863), while Gross et al. (2004) and Palva et al. (2005) suggest that lower frequency synchronisation is linked to attention. Similar to the debate above over early and late ERPs, the predictions from global workspace theories suggest that only late and sustained high frequency neural synchrony correlates with awareness. Other early forms of synchrony may contribute to this final state, but do not constitute the kind of synchrony associated with consciousness. Following on from the discussion above, much of this debate is due to the way in which different researchers characterise consciousness, and hence how they use different detection procedures to identify markers and measures of it.

As a potential way of establishing a detection procedure that is not based so strongly on a theoretical conception of consciousness or a commitment to a particular behavioural marker, Lamme (2006) offers his 'Neural Stance'. This is a 'bottom-up' way of identifying the neurophysiological markers of consciousness, which provides its own set of experimental protocols. Taking its lead from natural distinctions between types of neural activity, it does not rely on a behavioural marker of consciousness, or on differentiating *a priori* between the 'conscious' and 'unconscious' stages of neural processing in order to identify appropriate detection procedures. This 'Neural Stance' is discussed below in relation to local and global recurrent processing as a measure of consciousness.

5.3.3 Local vs. Global Recurrent Processing

Many researchers suggest that local recurrent processing is necessary for consciousness but that only *global* recurrent processing is sufficient for consciousness to occur. The evidence for the importance of global recurrent activity comes from many of the same paradigms investigating neural synchrony. Evidence from masking paradigms and the attentional blink suggest that brain activity in earlier local levels of processing is very similar for both 'seen' and 'unseen' trials. Divergence in activity only occurs later with widespread activity in prefrontal, parietal and temporal areas for 'seen' but not 'unseen' trials (see e.g. Dehaene et al. 2001; Del Cul et al. 2007; Sergent et al. 2005).

In contrast, Lamme 2003, 2004, 2006 (Lamme and Roelfsema 2000) has argued that local recurrent activity is sufficient in itself for conscious perception. His 'Neural Stance' is based on the recognition that many of the proposed neurophysiological measures of consciousness are determined by whatever behavioural criteria a researcher decides to use to operationalise consciousness. This will be based on their pre-theoretic commitments about what consciousness is (helpfully summarised by Lamme 2006, pp. 495–496). He states that: "This not only poses a problem for finding the 'true' NCC [neural correlate of consciousness, or appropriate measure of consciousness]; more serious is that, in this way, neuroscience will hardly fulfill its promise to get rid of the 'tedium of philosophers perpetually disagreeing with each other"' (Lamme 2006, p. 494).

He argues that in order to learn anything from neuroscience, distinctions in neural phenomena themselves must be capable of changing our preconceptions about what consciousness is. The distinction between feedforward and recurrent processing captures the differences between automatic and flexible processing, the ability to learn via synaptic plasticity, and the ability to integrate information, all plausible theoretically relevant distinctions between conscious and unconscious processing. The Neural Stance therefore comprises both a methodological point about the role of neuroscientific research in informing consciousness science, and a hypothesis that phenomenal consciousness is identical with recurrent processing.

Lamme argues that the Neural Stance offers an important alternative way of deciding what occurs in paradigms for which the behavioural evidence and standard experimental protocols are controversial. He argues that whatever a behavioural response might be, the neural evidence of the occurrence of recurrent processing is direct evidence for the occurrence of phenomenal consciousness. For example, research on inattentional blindness show that recurrent processing occurs in stimulus-specific visual areas when unattended objects are not reported (Scholte et al. 2006). Supèr et al. (2001) also found that stimulus-specific recurrent processing continued to occur in V1 of a macaque even though the monkey no longer 'reported' the presence of the stimulus due to a change in its report criterion.

Proponents of global workspace theories interpret these experiments as evidence that local recurrent processing is *not* sufficient for conscious perception, because it does not always suffice for report. However, Lamme states that the only important difference between cases of conscious and unconscious perception is a neural one: the presence of recurrent processing. The presence of recurrent processing shows that phenomenal consciousness is present, even for unattended and unreported stimuli. Reports and attention are experimental confounds of phenomenal consciousness using behavioural markers, and are best labelled as instances of access consciousness.

Given this hypothesis, Lamme claims that if the only important difference at the neural level is between feedforward and recurrent processing, then it should not matter what form the recurrent processing takes in order for it to be considered as a marker of consciousness. In particular, the location of recurrent processing should be irrelevant in distinguishing between conscious and unconscious processing: "... stimuli that evoke RP [recurrent processing] change your brain, while stimuli that evoke only feedforward activation have no lasting impact. It would be difficult to see why the involvement of the frontoparietal network would make such a difference" (Lamme 2006, p. 499). In this case, local recurrent processing in sensory areas is in itself sufficient for phenomenal consciousness.

This example again shows how experimental protocols can differ due to the research interests, theoretical commitments and methodologies of different research groups. Based on the Neural Stance, Lamme rejects the use of behavioural markers of consciousness and uses the difference between feedforward and recurrent processing to generate a set of experimental protocols. These identify early local recurrent processing with phenomenal consciousness, and later instances of global recurrent processing as instances of access consciousness.

5.3.4 Experimental Protocols: Production, Measurement and Detection Procedures

The differences in experimental protocols used to establish neurophysiological measures of consciousness described above can be summarised relative to the three features of Sullivan's framework. These are production procedures, or how stimuli are presented to subjects, measurement procedures, or what sort of response variables are measured, and detection procedures, or what kind of data analysis is carried out on the measurements. These are described below, followed by a discussion of whether the range of experimental protocols used in consciousness science can support the assumption that they all operationalise different aspects or types of a common phenomenon: consciousness.

As mentioned above, the production procedures across the paradigms used in consciousness science vary hugely. Some procedures make use of known attentional phenomena, such as the attentional blink (Del Cul et al. 2007; Sergent et al. 2005) and change blindness (Koivisto and Revonsuo 2003). Some include direct manipulations of response criteria by changing percentages of catch trials across sets of test trials (Supèr et al. 2001), most use visual stimuli but some use tactile stimuli (Palva et al. 2005), and all vary with regard to the specific stimuli and types (or lack) of masking used in the experimental paradigm. There is very little replication of stimulus presentation and task types across different paradigms. This shows that paradigms vary in at least the kind of sensory processing that is being tested, and sometimes in the relevance of attentional processes.

Second, measurement procedures differ across paradigms. For example, those supporting global workspace models of consciousness are explicit in their claims that attention, working memory and report are essential features of consciousness, so make use of subjective reports in a 'visibility' scale that gives a bimodal response distribution. In direct contrast to this, Lamme's neural stance implies that attention, working memory and report are experimental confounds, so reportability is not identified as a response variable. Others measure different types of responses such as detection, identification and two-alternative forced choice delayed match-to-sample tasks. Further, the way that neurophysiological data is measured also varies, with differences across paradigms in terms of the combination of imaging used, and the resolution of these measures (see discussion see Pins and Ffychte 2003). However, for present purposes the differences in measurement procedures across behavioural markers are sufficient for the claim that very different measures are taken.

Finally, paradigms vary in the detection procedures used. In terms of behavioural markers, responses can be assessed in terms of objective measures of sensitivity d' or response times (e.g. Melloni et al. 2007), or subjective visibility ratings (e.g. Del Cul et al. 2007), all giving very different ways of identifying the presence of consciousness. This also plays out at the level of analysing neurophysiological data. Dehaene and colleagues use a detection procedure that identifies the earliest bimodal patterns of brain activity across 'seen' and 'unseen' trials. In contrast, those endorsing the distinction between phenomenal and access consciousness (e.g. Lamme, Koivisto et al.), use detection procedures that identify the earliest neurophysiological differences between 'seen' and 'unseen' trials. They claim that early markers can be used to measure phenomenal consciousness, and later markers (such as those identified by Dehaene), as measures for access consciousness. Those using the same detection procedure (all differences between 'seen' and 'unseen' trials), but with a different conception of consciousness, view all markers picked out by the detection procedure as markers for consciousness, seen as a temporally extended process (e.g. Pins and Ffychte, Melloni et al., Palva et al.). In this case, there will be many markers for the many different functions that together allow subjects to process information and respond.

These different detection procedures and conceptions of consciousness all identify different sets of neurophysiological markers and measures of consciousness. While researchers often acknowledge these gross differences, they all argue that they are identifying the markers and measures of consciousness. This is the reason why integrative approaches are supposed to be appropriate; because they allow researchers to converge on an appropriate taxonomy and categorisation of the measures proposed so far.

5.4 Convergence: Where It's at

Having established what the differences are in production, measurement and detection protocols across measures of consciousness, and how deeply neurophysiological measures are tied to behavioural protocols and theoretical assumptions about consciousness, this section analyses where convergence between measures has been found or is likely to occur. Some instances of convergence between neurophysiological measures are illustrated above, while some strictly non-convergent results were found in earlier discussions of behavioural measures of consciousness. By considering all of these together, it is possible to map out the likely products of an integrative approach. Below, the three possible combinations of behavioural and neurophysiological measures are considered (BB, NN, and BN), where convergence can be found, and what can be learned from this.

5.4.1 Behavioural-Behavioural (BB)

Some comparisons across behavioural measures have been noted in earlier chapters, along with explanations of the presence or absence of convergence in particular cases. In general, convergence between certain types of behavioural measures is unlikely. For example, much was made of the distinction between subjective (report-based) and objective (performance-based) measures of consciousness and how they are measures of two fundamentally different phenomena; sensitivity and decision-making/report. Subjective measures are subject to response bias, so subjective and objective measures often diverge in experimental situations. Earlier chapters also showed how it was the very differences between Type 1 (responses about stimuli) and Type 2 (responses about responses to stimuli) subjective measures of consciousness that researchers used to build a measure of consciousness with.

Further, different ways of gathering subjective reports give divergent results. Sandberg et al. (2010) tested how consistent responses were across the Perceptual Awareness Scale (visibility ratings), confidence ratings and post-decision wagering (both variations on a Type 2 confidence rating task) in a masked visual identification task. They found that these measures differed in their response distributions (some bimodal, some not), and in the way they tracked task performance for different stimulus durations. Different subjective measures differ in the precise processes they measure and the factors that they are sensitive to, often giving different results. Among behavioural measures, both objective and subjective, there is little convergence, and much recognised divergence, between measures.

Further, as noted in Chap. 2, there are reasons to reject the equation between phenomenal and access consciousness with subjective and objective measures of awareness, respectively. The variations across different ways of gathering subjective reports found in the Sandberg et al. study provide further support for this claim. Objective measures assess sensitivity and subjective measures indicate the placement of response criteria, and are best described in these terms only.

5.4.2 Neurophysiological-Neurophysiological (NN)

There is clearly some convergence between neurophysiological measures as evidenced above. However, the basis of this convergence may be a rather uninteresting one. Within a research group, a range of neurophysiological markers are usually found to correlate with consciousness. Those who use attentional blink paradigms and search for bimodal patterns of activity identify late ERPs, late and sustained neural synchrony and global recurrent processing as markers of consciousness. Those using different experimental protocols find quite different results. So, while there is some convergence among neurophysiological markers for the *same* experimental protocols, there is little or none *across* protocols.

This is not particularly surprising. Neurophysiological markers are markers of the brain activity that occurs during a specific experimental paradigm, typically in order to generate a particular behavioural response. Different ways of measuring this activity are all based on the presence of neural synchrony. So, neural synchrony can be investigated in terms of its frequency range, duration (EEG) and location (fMRI), it can be investigated in terms of specific task-related or context independent waveforms using EEG (ERPs), or it can be investigated in relation to its presence across neural hierarchies using a mix of EEG and fMRI (recurrent processing). Since many of these investigations overlap, there is likely to be convergence within specific experimental paradigms across these markers.

However, for different paradigms involving different tasks, different kinds of brain activity will occur. Using different detection procedures, different stages of processing will be selected. Those who select late neural activity as a marker of consciousness in an attentionally demanding paradigm are assessing different processes, and in a different ways, than those who select early neural activity as a measure of consciousness using paradigms that have low attentional demands. In this case, neurophysiological measures across experimental protocols are unlikely to converge.

5.4.3 Behavioural-Neurophysiological (BN)

As suggested above, there are significant differences between behavioural protocols that prevent meaningful convergence among behavioural measures. With the exception of local RP, neurophysiological measures are all dependent on behavioural markers, so neurophysiological measures are simply measures of whatever phenomena are operationalised by a behavioural measure. This obvious dependence of neurophysiological on behavioural measures presents a problem for those seeking any interesting convergence between the two. As Craver (2007) remarks: "If the fields and techniques were *not* largely autonomous, if the results of one *could* be translated into the results of the other, then they would not provide *independent* evidence..." (p. 240, original italics).

The dependence of neurophysiological work on behavioural operationalisations of consciousness means that they do not provide independent evidence. This precludes them from being the subject of any insightful or productive integrative techniques. For example, neurophysiological measures will obviously correlate with the behavioural measures they are based on, but in this case they are simply two different ways of assessing exactly the same phenomenon (see also Lamme's criticism of this method above).

Any interesting convergence can only emerge between *different* experimental protocols. However, as seen above, if two research labs differ in the task or detection procedure used to operationalise consciousness, they typically find quite different neural markers. A neurophysiological measure found in a paradigm using bimodal subjective report in an attentional blink paradigm, will not correlate with behavioural and neurophysiological markers of forced choice performance in a simple detection task. A neurophysiological marker of phenomenal consciousness (e.g. pre-attentional processes) will not correlate with behavioural or neurophysiological markers of Type 2 confidence ratings. The same problems that prevent successful integration of behavioural measures play out yet again at the neurophysiological level.

This is because, as Lamme points out, neurophysiological measures are typically based entirely on the same assumptions and protocols used to support behavioural measures, yet behavioural measures are themselves very divergent measures. Indeed, as seen from the discussions above, there seems to be little convergence across current experimental protocols between behavioural and neurophysiological measures. It would also be strange to expect similarities in brain activity across attentional and preattentional processing, or perceptual and metacognitive monitoring, or simply across different stages of different perceptual tasks, to emerge.

5.4.4 Learning from the Lack of Convergence

Far from seeing the wide range of theories, behavioural measures, and dependent neurophysiological measures of consciousness as impediments to convergence, Seth et al. (2008) suggest that they can be used to inform different theories and measures of consciousness: "Just as theoretical positions conflict with one another, conflicts among measures can be expected and, in many cases, have been observed. These conflicts can guide further experiments and theoretical refinements" (pp. 319–320).

They suggest that the differences between Type 2 confidence ratings that involve wagering and other behavioural responses can be used to establish if they assess different processes. Further, they suggest that the kind of neural activity sufficient for reportability conflicts with simple theories about the kind of information processing necessary for report, promoting the development of new ways of measuring and modeling neural interactions. In itself, these are valuable things to learn from a comparative, integrative approach. Discovering distinctions in what different measures appear to assess, and improving models and theories about what kinds of processing is necessary for certain behavioural responses, is clearly progress. Shea and Bayne (2010) also state that there seems to be convergence for measures of 'determinates' of phenomenal consciousness, though these determinates may be rather different from each other (e.g. visual/auditory experience).

However, it is not a straightforward move to claim that these sets of convergent measures should be categorised as different types of consciousness. Afterall, these sets of convergent measures are converging around rather different phenomena (reportability, Type 2 confidence ratings, pre-attentional processing). That is, it is not clear whether these clusters of convergence are clusters related to a common phenomenon (consciousness), or if they should be characterised as clusters relating to different phenomena altogether. The final section of this chapter deals with this question, which is developed further in Chap. 6.

5.5 Convergence Towards What?

There are a number of reasons why a group of apparently disparate measures can be usefully viewed as measures of different types of a common phenomenon. A rather obvious reason is that the measures all give similar patterns of results under a range of experimental manipulations. This can be as simple as the measures all changing in a correlated way with longer stimulus duration or practice (e.g. accuracy rates going up and response times going down). The mere fact that there are a set of manipulations that tend to affect a group of phenomena in the same way is some reason to consider them as types of the same common phenomenon.

For example, despite the recognised taxonomy of memory into a range of different phenomena, with each relying on different mechanisms (see e.g. Bechtel 2008 for discussion of the history of memory research), there are a range of manipulations that tend to affect all of them in similar ways. If the presentation of a target stimulus is compromised in some way, for example if it is presented for a short duration, or is heavily masked, then subjects tend to perform worse on measures of memory. If there is a long delay between the presentation of the target stimulus and the test, then subjects again tend to perform worse on measures of memory. Though different types of memory are constituted by different mechanisms in different parts of the brain, they all support these two abstract stages of information processing (encoding, storage, and a third being retrieval).

Based on the ways that measures of memory respond to particular experimental manipulations, memory can be given a functional description, along with some factors that affect its function. Common functional descriptions also allow for generalisations and predictions to be made across the range of phenomena, for a range of manipulations. Using these two related ideas, it is possible to examine what kinds of similarities there are across the behavioural and neurophysiological measures discussed so far, to see if there is sufficient convergence to warrant the inference to a common phenomenon.

Of course, common functions, predictions and generalisations can be established for specific sets of convergent measures, as is clear from the earlier discussions. For example, behavioural and neurophysiological measures converge for paradigms like the attentional blink, using subjective measures of performance. Both reports and the P3 ERP are affected in convergent ways for the same manipulations (in the attentional blink subjects fail the task, and do not exhibit the P3 waveform). If consciousness is associated with report and the P3 waveform, its function can be described as allowing for the global availability and integration of information, resulting in reports.

However, whatever predictions, generalisations and functionable descriptions can be generated for these cases of convergence can also easily be described in terms of the specific sensory or cognitive phenomena that are used to operationalise consciousness (e.g. attention and report). For example, in attentional blink paradigms, this includes subjective report, global recurrent processing and late sustained gamma band neural synchrony as measures for attention, working memory and decision-making, 'detected' using bimodal response distributions. Alternatively, forced choice responses, earlier, transient and lower bandwidth neural frequency and local recurrent processing may be a convergent set of measures for the range of capacities operationalised in forced choice word categorisation paradigms, using the detection procedure of looking for the first significant differences in 'seen' vs. 'unseen' trials.

What is necessary for measures of consciousness to be characterised as measures of different types of a single common phenomenon is that there is a common functional description shared by all (and only) the measured phenomena, and that reliable predictions and broad generalizations can be made across all (and only) those phenomena. One similarity is that reduced stimulus presentation times and increased masking reduces the ability of subjects to perform well across all behavioural protocols, and thus across all behavioural measures, that are used in consciousness science. Neurophysiological measures will track these behavioural measures so are also affected in a similar way by these manipulations. Without attention, behavioural performance also tends to decrease, but the use of different detection procedures for neurophysiological measures means that this is not always tracked at the neural level (e.g. by explicitly looking for pre- or post-attentional processing).

It is difficult however to find further similarities in the way that all of the measures discussed above are affected by experimental manipulations (and these may not even be true for Lamme's proposal of local RP). The properties of some measures, that they pick out instances of control, flexible information processing, reportability, and so on, are not found across *all* measures, so cannot be used to give a functional description that covers them all. Perhaps an appropriate functional description could be that conscious perception enables 'better' information processing (deeper, more reliable information processing).

However, a problem with this is that it is unlikely to be unique to the phenomena assessed by measures of consciousness. For example, it is also true that manipulations of stimulus presentation times, masking and attention affect subjects' performance in a similar way in (putatively) *unconscious* perception paradigms. Subjects exhibit effects of deeper and more reliable processing for longer stimulus durations, weaker masking, and stimuli presented in attended areas even when stimuli are presented under the threshold for conscious perception, whether taken as reportability or $d' = 0$ (e.g. see Kouider and Dehaene 2007). Given that manipulations of stimulus presentation conditions affect information processing in a similar way for purported instances of both conscious and unconscious perception, there seems to be little that is unique to all measures of consciousness in functional terms. In turn, this strongly suggests that there are no unique predictions and generalisations that can be made across all (and only) measures of consciousness. Measures of consciousness seem, at best, to be an ill-defined subset of measures that assess information processing.

Indeed, as suggested above, different measures often assess fundamentally different processes that have little or nothing in common. Measures of local recurrent processing certainly seem to share little in functional terms with measures of attention or decision-making. This suggests further that the assumption that these measures all assess different aspects or types of a single phenomenon is simply not compatible with the scientific evidence to date. The lack of a common unique function, derived from a set of common properties, also ensures that there are no unique cross-measure predictions or generalisations that can be made.

In fact, prior to consciousness science lumping these phenomena together and relabelling them as instances of graded or different types of consciousness, descriptions that made reference to specific sensory, cognitive or neural phenomena were perfectly adequate. These include different kinds of perception, different types of responses, different types of attention, different types of neural processing, and so on. Crucially, in line with earlier chapters, these descriptions and categories of phenomena make reference to the experimental manipulations used to investigate them. They support generalisations and predictions, and promote clear communication across research groups. The labels offered by consciousness science are unclear and leave out vital information found in the more specific descriptions using the vocabulary of the cognitive and neurosciences.

Recognising this, Hulme et al. (2008) proposed that until a definition of consciousness can be agreed on, talk of consciousness should be dropped and the cognitive abilities that researchers are actually assessing in experimental paradigms should be made explicit. As part of this program, Hulme et al. have isolated correlates for different aspects of report generation, including stimulus-specific processing, decision-making, and verbal report:

> [...] it is more appropriate for neuroscience to restrict the interpretations of NCC [neural correlates of consciousness—neurophysiological markers] experiments to the operational marker used; not what they interpret it as standing for. Even when interpreting NCC data purely in terms of the operational marker, it is crucial to delineate the neural correlates of its components. (Hulme et al. 2008, p. 1609)

There seems little scope, outside appeals to intuition and the determination to preserve concepts of consciousness, to argue that the neurophysiological measures above, and the behavioural measures discussed in earlier chapters, are best characterised as measures of different types or aspects of a common phenomenon. The lack of a set of uniquely identifying common properties, the implications of a lack of a common function, the failure to generate unique predictions or generalisations across all measures of consciousness, the absence of any links to the experimental paradigms used to investigate this concept, and the impact on communication across research groups, provide many reasons for avoiding using the concept of consciousness. Instead, by referring to specific sensory, cognitive and neural phenomena that are actually investigated in experimental paradigms, these problems do not arise. While convergence towards measures of attention, or decision making or early stages of sensory processing is likely to occur (as evidenced above), there is no empirical or theoretical support for interpreting this as convergence among measures of the common phenomenon of consciousness.

5.6 Conclusion

This chapter had three aims. The first was to suggest that certain types of variation across experimental paradigms can impede successful integration and convergence, and to make explicit, using Sullivan's framework of experimental protocols, the different ways in which consciousness is operationalised and measured. The second section identified where convergence among behavioural and neurophysiological measures occurs, and explained why convergence was found among some measures and others. The final section tried to establish whether or not the convergence found among putative measures of consciousness is sufficient to justify the assumption that they are all measures of different types of a common phenomenon, or whether they are best viewed as measures of a range of distinct phenomena. In this second case, some suggestions were made as to what these phenomena are.

The claims made here go beyond Sullivan's claims about the lack of convergence in neuroscientific research. She argues that there are significant differences in experimental protocols used to operationalise 'the same' neuroscientific phenomenon, and that these ensure that they instead operationalise quite different phenomena. Due to these differences, integrated models and explanations of neuroscientific phenomena are unlikely to emerge from current research practices. However, this does not prevent convergence happening in the future. There is at least some agreement over what LTP and social recognition memory are. If researchers establish some common experimental protocols with an accepted range of variation across them, then integrated and convergent models and explanations can emerge. Sullivan's arguments are based on current experimental practices, so her claims are limited to current research.

In consciousness science there are also significant differences across production, measurement and detection procedures in experimental protocols for investigating consciousness. It is clear from the above that local clusters of convergence between sets of measures can be observed. However, the convergence that exists within consciousness science can be identified as the product of investigating similar sets of sensory, cognitive or neural phenomena across groups of experimental paradigms. These distinct phenomena are already the subjects of investigation in cognitive and neuroscience. Further, there seem to be no reasons for viewing these phenomena as different types of a common phenomenon.

This is where the claims made here differ from Sullivan's claims about neuroscience. While convergence within current neuroscientific research into LTP and social recognition memory seems unlikely now, it is possible in the future. In contrast, convergence towards measures of consciousness is not only unlikely now, but there are strong reasons to argue why it should not be sought. Measures of consciousness do not identify a set of phenomena that share a unique common function, or that can support cross-measure predictions or generalisations. Categorising already well-known phenomena as types or aspects of consciousness is not a productive theoretical move. Now gathered under the banner of consciousness research, cognitive and neuroscientific research is best left as it was. This would allow it to continue to establish distinctions between phenomena based on real divergence and convergence, and to label these phenomena in a clear and methodologically appropriate way.

An extension of these ideas can be found in the next chapter. In discussing the ways of identifying common phenomena above, some appeals were made to common features and common functions. Boyd (1991) has proposed that natural groupings of phenomena that support generalisations and predictions, scientific kinds, can be identified through clusters of co-occurring properties generated by a common mechanism (or set of mechanisms). Therefore a way of pursuing the claim that the phenomena investigated in the science of consciousness are not different types of consciousness, but a range of different phenomena altogether, is to consider proposed mechanisms of consciousness. It will be suggested that these mechanisms of consciousness are badly demarcated, are targeted at more narrow or different phenomena than their proponents aim at, and are most naturally described in terms of specific information processing or cognitive functions. The methodological problems associated with failures to identify anything like a good candidate mechanism for consciousness again suggest that 'consciousness' is not a viable target for scientific research.

References

Baars, B. J. (1997). In the theatre of consciousness: Global workspace theory, a rigorous scientific theory of consciousness. *Journal of Consciousness Studies, 4,* 292–309.

Bechtel, W. (2008). *Mental mechanisms: Philosophical perspectives on cognitive neuroscience.* London: Routledge.

Bentin, S., McCarthy, G., & Wood, C. C. (1985). Event-related potentials, lexical decision and semantic priming. *Electroencephalography and Clinical Neurophysiology, 60,* 43–355.

Bickle, J. (2006). Reducing mind to molecular pathways: Explicating the reductionism implicit in current cellular and molecular neuroscience. *Synthese, 151,* 411–434.

Boyd, R. (1991). Realism, anti-foundationalism and the enthusiasm for natural kinds. *Philosophical Studies, 61,* 127–148.

Collins, A. M., & Loftus, E. F. (1975). A spreading-activation theory of semantic processing. *Psychological Review, 82,* 407–428.

Craver, C. F. (2007). *Explaining the brain: Mechanisms and the mosaic unity of neuroscience.* Oxford: Clarendon Press.

Crick, F., & Koch, C. (1990) Towards a neurobiological theory of consciousness. *Seminars in the Neurosciences, 2,* 263–275.

Dehaene, S., Naccache, L., Le Clec' H. G., Koechlin, E., Mueller, M., Dehaene-Lambertz, G., Moortele, P. F. van de, & Le Bihan, D. (1998). Imaging unconscious semantic priming. *Nature, 395,* 597–600.

Dehaene, S., Naccache, L., Cohen, L., Le Bihan, D., Mangin, J.-F., Poline, J.-B., & Rivière, D. (2001). Cerebral mechanisms of word masking and unconscious repetition priming. *Nature Neuroscience, 4,* 752–758.

Dehaene, S., Changeux, J. P., Naccache, L., Sackur, J., & Sergent, C. (2006). Conscious, preconscious, and subliminal processing: A testable taxonomy. *Trends in Cognitive Sciences, 100,* 204–211.

Del Cul, A., Baillet, S., & Dehaene, S. (2007). Brain dynamics underlying the nonlinear threshold for access to consciousness. *PLoS Biology, 5*(10), e260.

Fries, P., Schröder, J.-H., Roelfsema, P. R., Singer, W., & Engel, A. K. (2002). Oscillatory neuronal synchronization in primary visual cortex as a correlate of stimulus selection. *The Journal of Neuroscience, 22,* 3739–3754.

Gross, J., Schmitz, F., Schnitzler, I., Kessler, K., Shapiro, K., Hommel, B., & Schnitzler, A. (2004). Modulation of long-range neural synchrony reflects temporal limitations of visual attention in humans. *Proceedings of the National Academy of Sciences, 101,* 13050–13055.

Holcomb, P. J. (1993). Semantic priming and stimulus degradation: Implications for the role of the N400 in language processing. *Psychophysiology, 30,* 47–61.

Hulme, O. J., Friston, K. F., & Zeki, S. (2008). Neural correlates of stimulus reportability. *Journal of Cognitive Neuroscience, 21,* 1602–1610.

Koivisto, M., & Revonsuo, A. (2003). An ERP study of change detection, change blindness, and visual awareness. *Psychophysiology, 40,* 423–429.

Koivisto, M., Revonsuo, A., & Salminen, N. (2005). Independence of visual awareness from attention at early processing stages. *NeuroReport, 16,* 817–821.

Koivisto, M., Revonsuo, A., & Lehtonen, M. (2006). Independence of visual awareness from the scope of attention: An electrophysiological study. *Cerebral Cortex, 16,* 415–424.

Koivisto, M., Railo, H., Revonsuo, A., Vani, S., & Salminen-Vaparant, N. (2011). Recurrent processing in V1/V2 contributes to categorization of natural scenes. *The Journal of Neuroscience, 31,* 2488–2492.

Kouider, S., & Dehaene, S. (2007). Levels of processing during non-conscious perception: A critical review of visual masking. *Philosophical transactions of the Royal Society B, 362,* 857–875.

Lamme, V. A. F. (2003). Why visual attention and awareness are different. *Trends in Cognitive Sciences, 7,* 12–18.

Lamme, V. A. F. (2004). Separate neural definitions of visual consciousness and visual attention; a case for phenomenal awareness. *Neural Networks, 17,* 861–872.

Lamme, V. A. F. (2006). Towards a true neural stance on consciousness. *Trends in Cognitive Sciences, 10,* 494–501.

Lamme, V. A. F., & Roelfsema, P. R. (2000). The distinct modes of vision offered by feedforward and recurrent processing. *Trends in Neuroscience, 23,* 571–579.

Melloni, L., Molina, C., Pena, M., Torres, D., Singer, W., & Rodriguez, E. (2007). Synchronization of neural activity across cortical areas: Correlates with conscious perception. *The Journal of Neuroscience, 27,* 2858–2865.

Palva, S., Linkenkaer-Hansen, K., Näätänen, R., & Plave, J. M. (2005). Early neural correlates of conscious somatosensory perception. *The Journal of Neuroscience, 25,* 5248–5258.

Perea, M., & Gotor, A. (1997). Associative and semantic priming effects occur at very short stimulus-onset asynchronies in lexical decision and naming. *Cognition, 62,* 223–240.

Pins, D., & Ffychte, D. (2003). The neural correlates of conscious vision. *Cerebral Cortex, 13,* 461–474.

Raymond J. E., Shapiro K. L., & Arnell K. M. (1992). Temporary suppression of visual processing in an RSVP task: An attentional blink? *Journal of Experimental Psychology. Human Perception and Performance, 18,* 849–860.

Sandberg, K., Timmermans, B., Overgaard, M., & Cleeremans, A. (2010). Measuring consciousness: Is one measure better than the other? *Consciousness and Cognition, 19,* 1069–1078.

Scholte, H. S., Witteveen, S. C., Spekreijse, H., & Lamme, V. A. F. (2006). The influence of inattention on the neural correlates of scene segmentation. *Brain Research, 1076,* 106–115.

Scholte, H. S., Jolij, J., Fahrenfort, J. J., & Lamme, V. A. F. (2008). Feedforward and recurrent processing in scene segmentation: Electroencephalography and functional magnetic resonance imaging. *Journal of Cognitive Neuroscience, 20,* 2097–2109.

Sergent, C., Baillet, S., & Dehaene, S. (2005). Timing of the brain events underlying access to consciousness during the attentional blink. *Nature Neuroscience, 8,* 1391–1400.

Seth, A. K., Dienes, Z., Cleeremans, A., Overgaard, M., & Pessoa, L. (2008). Measuring consciousness: Relating behavioural and neurophysiological approaches. *Trends in Cognitive Sciences, 12,* 314–321.

Shea, N., & Bayne, T. (2010). The vegetative state and the science of consciousness. *British Journal for the Philosophy of Science, 61,* 459–484.

Sullivan, J. (2009). The multiplicity of experimental protocols: A challenge to reductionist and non-reductionist models of the unity of neuroscience. *Synthese, 167,* 511–539.

Supèr, H., Spekreijse, H., & Lamme, V. A. F. (2001). Two distinct modes of sensory processing observed in monkey primary visual cortex (V1). *Nature Neuroscience, 4,* 304–310.

Uhlhaas, P. J., & Singer, W. (2006). Neural synchrony in brain disorders: Relevance for cognitive dysfunctions and pathophysiology. *Neuron, 52,* 155–168.

Uhlhaas, P. J., Pipa, G., Lima, B., Melloni, L., Neuenschwander, S., Nikolic, D., & Singer, W. (2009). Neural synchrony in cortical networks: History, concept and current status. *Frontiers in Integrative Neuroscience, 3,* 1–19.

Visser, T. A. W., & Merikle, P. M. (1999). Conscious and unconscious processes: The effects of motivation. *Consciousness and Cognition, 8,* 94–113.

Chapter 6
Conscious Mechanisms and Scientific Kinds

6.1 Introduction

This Chapter extends the arguments found in Chap. 5 by trying to establish what sort of things are really being investigated in consciousness science. One way of doing this is to look for the phenomena about which reliable predictions and broad generalisations can be made; sometimes referred to as scientific kinds. These things or phenomena are what a science is really about, as predicting and generalising are arguably what successful science does. The obvious claim is that the kinds investigated in consciousness science are kinds of consciousness. However, as suggested by previous chapters, this claim is not necessarily supported by the empirical investigations that occur in consciousness science.

The previous chapter focused on the attempt to find clusters of measures of consciousness that were supposed to indicate the presence of different types (kinds) of consciousness. Expanding on this idea, Boyd 1991 (Kornblith 1993) has argued that scientific kinds can be identified as commonly co-occurring clusters of properties, and the mechanisms that produce them. This chapter will analyse proposed mechanisms of consciousness to see whether they are well-demarcated, and whether they produce the phenomena that their proponents claim they do. This will not only show if standard practices for identifying mechanisms are used in consciousness science, but can also be used to identify the kinds that consciousness science is, and is not, about.

As seen in the quotation from Crick and Koch in the first chapter, researchers tend to assume that consciousness stems from a set of common mechanisms: "The second assumption is tentative: that all the different aspects of consciousness, for example pain and visual awareness, employ a basic common mechanism or perhaps a few such mechanisms. . . " (Crick and Koch 1990, pp. 263–264). This assumption is still held by researchers, including those discussed in this chapter who identify two very different mechanisms of consciousness. These are the mechanisms proposed in Dehaene's Global Neuronal Workspace Theory (2001, 2006) and Lamme's Neural Stance (2006).

Following an outline of Boyd's account of scientific kinds and a discussion of Craver's (2007) account of how mechanisms are demarcated in practice, these two mechanisms will be analysed to see if they constitute scientific kinds of

E. Irvine, *Consciousness as a Scientific Concept,* Studies in Brain and Mind 5, DOI 10.1007/978-94-007-5173-6_6,

consciousness. This analysis includes a discussion of how well Craver's account of the demarcation of mechanisms applies to Global Neuronal Workspace Theory, and problems with the internal consistency of the way in which Lamme identifies a mechanism of consciousness. Finally, it will be questioned if the mechanisms actually produce the phenomena that Dehaene and Lamme seek to explain, or something else entirely. The scientific kinds present in consciousness science can then be identified, thus shedding more light on whether or not 'consciousness' is a viable target for scientific investigation.

6.2 Property Clusters and Scientific Kinds

Boyd's (1991) Homeostatic Property Cluster theory of scientific kinds states that they can be defined as commonly co-occurring clusters of properties whose common co-occurrence is the product of an underlying mechanism (see also Boyd 1989, 1997; Kornblith 1993). This is in contrast with traditional accounts of natural kinds in which there has to be a set of necessary and sufficient conditions for kind membership, which often fail to capture biological kinds. For example, there are no necessary and sufficient conditions for something being a dog; while most dogs are hairy, four-legged carnivorous pack animals, some are not (a three legged dog is a still a dog). At the genetic level, while much might be shared across all dogs, there is clearly a wide range of variation in genotypes, and genotypes can be expressed differently depending on environmental factors.

Despite this variation however there are a wide range of predictions and generalisations that can be made about dogs, and biology plausibly does investigate the properties of scientific kinds. So, Boyd argues that a more fitting way of describing scientific kinds seems to be in terms of prediction and generalisation. By identifying Homeostatic Property Clusters (HPCs) that consist of commonly co-occurring clusters of properties realised by a mechanism, reliable predictions and broad generalisations about the members of a scientific kind can be stated across a relevant range of background conditions. HPCs are a way of identifying scientific kinds that are grounded in the goals and the products of scientific practice: "Both the property-cluster form of such definitions and the associated indeterminacy are dictated by the fundamental epistemic task of employing categories which correspond to inductively and explanatorily relevant causal structures" (Boyd 1991, pp. 141–142).

Shea and Bayne (2010) have also used this idea of a scientific kind as a way of making progress on the taxonomisation and measurement of consciousness. They suggest that by finding behavioural and neurophysiological measures that correlate well with each other, it may be safe to infer that there is some common property, or some kind of consciousness, that all the measures in this cluster assess. In this case, a traditional measure of consciousness, such as reportability, may form part of a much larger cluster of tests, all of which assess a common kind. This is particularly useful for assessing subjects for whom the normal behavioural measures are inappropriate, such as vegetative state or locked in patients.

The previous chapter took a rather more skeptical stance towards the success of this method, where it was argued that while there are indeed property clusters to be found, they are not best described as forming different kinds of consciousness. However, the point remains that whatever the scientific kinds are in consciousness science, they will be discovered through scientific means. This chapter follows a different strategy to that found earlier and focuses on the mechanistic aspect of Boyd's HPC account. Investigating whether candidate mechanisms for consciousness are well-demarcated, and whether they produce the phenomena that are assumed to, offers an alternative way of assessing the scientific kinds under investigation in consciousness science.

It must be noted that using the HPC account to identify kinds does not generate a single, static categorization of phenomena, but potentially a range of overlapping characterisations (and kinds) relating to the same target phenomenon (Boyd 1999; Craver 2009). That is, what counts as sufficient 'clustering', or a part of a mechanism, will differ according to particular research goals. However, the HPC account can at least provide reasonable boundary conditions for the properties and mechanistic parts for a scientific kind, given a particular research context. In this way, the HPC account has been used in biology as a way of identifying scientific kinds that pick out real, rather than artificial property clusters, in virtue of a well-demarcated common mechanism or set of related mechanisms at work (e.g. Griffiths 1997 on emotion; Wilson 1999 on species).

6.3 Mechanisms and Distinctions in Consciousness Science

Mechanisms are described by Machamer et al. (2000) as "entities and activities organized such that they are productive of regular changes from start or set-up to finish or termination conditions" (p. 3). Mechanisms are organised groups of spatially and temporally distributed working parts that, in the right background conditions and with a standard input, together regularly produce a change or product. One of the significant questions about mechanisms is how their background conditions and constitutive components should be demarcated. In consciousness science, this question is phrased in terms of the 'total' (inclusive of background conditions) and 'core' (exclusive of background conditions) neural correlates of consciousness (NCCs). As Chalmers (2000) notes, the total NCC may be the entire brain. A core NCC on the other hand could be much smaller:

> A total NCC builds in everything and thus automatically suffices for the corresponding conscious states. A core NCC, on the other hand, contains only the 'core' processes that correlate with consciousness. The rest of the total NCC will be relegated to some sort of background conditions, required for the correct functioning of the core... The question is then how to distinguish the core from the background. (Chalmers 2000, p. 21)

The distinction between core and total NCCs is often related to the distinction drawn between correlates/mechanisms that provide the contents of consciousness (content NCCs), and those that support being conscious at all (state NCCs). However, this distinction is not necessarily a useful one, or one that is often invoked in empirical work. State consciousness is even referred to as a background state by Chalmers, and many of those working on 'state' consciousness refer to it in more specific terms, for example in terms of the differences between neural activity observed during alertness, different phases of sleep, coma, and permanently vegetative state, and so on (Laureys 2005; Laureys et al. 2004; Sadaghiani et al. 2010).

Instead, what many researchers are interested in is how content provided by sensory areas comes to be conscious (however consciousness is operationalised). This means that instead of the interesting distinction being between mechanisms of awakeness/alertness, and mechanisms for the contents of consciousness, the distinction is instead stated in terms of the mechanism for processing content and the mechanism that makes that specific content conscious. However, this distinction is again not universally supported or invoked across all scientific discussions of consciousness. These mechanisms may overlap and make use of some of the same components.

For example, workspace theories state that content is processed locally and becomes conscious when activity in local areas is attentionally amplified. In this case the components involved in processing content are also found in the mechanism for making that content conscious. Lamme's Neural Stance states that the components that generate the contents of consciousness and those that ensure that it becomes conscious are exactly the same (whether they are engaged in recurrent processing or not makes the difference). Problems with the distinction between mechanisms for content and consciousness of that content will be elaborated on in later sections.

However, even if state vs. content consciousness, and content vs. consciousness of content, are not necessarily useful distinctions to draw, the basic problem of how to distinguish between background conditions and core components plays a crucial role in current debates about the mechanism for consciousness. In investigating any large and complex system a set of recognised experimental techniques and criteria are required to isolate subsystems that generate particular phenomena. In relation to consciousness science in particular, Chalmers (2000) discusses a range of experimental methods (e.g. brain lesions, brain stimulation) and notes which are the more useful ones to identify core and background components of NCCs.

Craver (2007) offers a more general framework to identify constituent and background parts of mechanisms (and correlates), derived from considering experimental practices in neuroscience more generally. While there is a significant degree of overlap between Craver's framework and Chalmers discussions (e.g. the utility of stimulation studies, potential problems with lesion experiments), Craver offers a more structured, more general, and better developed account of the demarcation of mechanisms. This chapter will therefore make use of Craver's account, as it provides a framework that was developed independently of the particular debates in consciousness science.

Craver's account of the demarcation of mechanisms offers a practical account of how background conditions are separated from constitutive components of a system, based on assessing experimental techniques and the application of local knowledge or pragmatic factors to the investigation of a particular phenomenon. It is not an account that is supposed to provide clear-cut answers in all situations, but suggests how neuroscientists do, and how they should, demarcate mechanisms. This makes it possible to test whether consciousness science follows standard methodologies rather than its own methods, which may be, (as seen in earlier chapters), driven more by pre-theoretical assumptions and intuitions than good scientific practice. This account is outlined below.

6.3.1 Craver's Account of Mutual Manipulability

Craver (2007) argues that a part is a constitutively relevant component of a mechanism for a specified phenomenon, rather than a background condition for it, if the component and the phenomena are mutually manipulable under ideal interventions. The idea is that a component is part of a mechanism if, by wiggling the phenomenon you wiggle the component, and by wiggling the component you wiggle the phenomenon (see e.g. Craver 2007, p. 153). The wiggles should conform to conditions of ideal interventions such that the wiggles are directly causally related, e.g. that the wiggles are not both caused by a common cause, or that the effect-wiggle is caused indirectly by something else, and so on (Craver 2007, pp. 139–160 for more details). Further conditions are added to account for cases of redundancy, recovery and reorganisation in complex systems (pp. 156–157).

Craver states four criteria through which to assess degrees of mutual manipulability (mutual wiggliness) to test whether a candidate part is a constitutively relevant component of a mechanism for a particular phenomenon. The first criterion is that when the component is altered it must alter the phenomenon in question, and when the component is stimulated it must stimulate the occurrence of the phenomenon. The second criterion is that changes in the phenomenon change the component in question. Two other criteria for demarcating constitutively relevant components of a mechanism from background conditions are that changes in the component produce both specific and subtle changes in the phenomenon. Just how specific and subtle these changes need to be is debatable, but in actual cases it may be more obvious how these criteria play out (see Craver 2007, pp. 139–162 for more details). The way these criteria are applied is best described through a worked example, in this case in determining the components of the mechanism that enables motion detection, now commonly identified with activity in area V5/MT of visual cortex. This example will also be used later to support one of the key claims of this chapter.

The way in which the mechanism of motion detection has been investigated is through a combination of experimental techniques, invoking the notion of mutual manipulability above. For example, some of the earliest ways of investigating neural mechanisms were through lesion studies in animals, and investigating the effects of localised head injuries in humans. By lesioning area V5/MT, primates exhibit marked deficits in motion detection (Newsome and Paré 1988; Pasternak and Merigan 1994), and studies in humans also suggest that V5/MT damage affects motion detection (Hess et al. 1989; Nawrot and Rizzo 1988). This evidence addresses part of the first criterion of mutual manipulability that changes in the component change the phenomenon in question.

The other part of the first criterion is that stimulations of V5/MT stimulate motion detection. Indeed, electrical stimulation of V5 biases responses towards moving stimuli in primates (Nichols and Newsome 2002; Salzman et al. 1992), and can enhance motion perception in humans (Antal et al. 2004). In this case, V5/MT fulfils the first criterion of mutual manipulability, and can be considered as a candidate component for the mechanism for motion detection.

The second criterion is that changes in motion detection change the activity of area V5/MT. One way of doing this is simply to look for correlations between activity in V5/MT and performance in a range of motion detection tasks. By controlling for activity related to other stimulus features, it can indeed be shown that moving stimuli evoke activity in V5/MT and correlate with reports of motion in such a way that this second criterion is fulfilled (Zeki 1990; He at al. 1998). That is, differences in motion detection ability can be tracked in differences in the activity of V5/MT.

Of course, other brain areas will also satisfy these first two criteria to a greater or lesser extent. Stimulating earlier visual areas (e.g. the retina) will produce changes in motion detection and changes in motion detection can be tracked by changes in these areas. However, the last two criteria of mutual manipulability state that mutual manipulations must be subtle and specific. Stimulations of earlier visual processing areas cause changes not only in motion detection ability, but in other abilities too, including shape, colour, and object identification. However, stimulation of V5/MT seems to be fairly limited to stimulations of motion detection. Further, changes in motion detection are most accurately tracked through changes in V5/MT. In this case, V5/MT satisfies the specificity criterion.

It is also the case that the effects of stimulations of V5/MT, and changes in V5/MT given changes in motion detection, are fairly subtle. Stimulations of direction-tuned neurons produce shifts in reports of direction of motion, and changes in reported direction of motion can be tracked by changes in activity for groups of neurons. This degree of subtlety cannot be found in stimulations and changes in other visual areas. Therefore, area V5/MT fulfils all four conditions for being a component for the mechanism of motion detection. Accordingly, area V5/MT has been identified as the main component in the mechanism for motion detection, with other visual areas treated as more or less important background components.

This account of the demarcation of mechanisms will be used to assess the claims made in Deheane's Global Neuronal Workspace Theory (GNWT) about the mechanism of consciousness. The account can also be used to assess the distinction between correlates/mechanisms of conscious content, and correlates/mechanisms of consciousness of that content, that are often invoked in discussing consciousness. A different approach, based on assessing the internal consistency of claims made, will be applied to the mechanism proposed under Lamme's 'Neural Stance'.

Importantly, GNWT and the Neural Stance exemplify two different ways of identifying mechanisms, as described in Bechtel (2008). GNWT focuses on identifying core behavioural properties and functions of the phenomena in question, and from here searching for the mechanism that produces these properties. This can be described as a top-down approach, in which researchers specify a (higher-level) target phenomenon and then search for the (lower-level) mechanism that produces it. Lamme's approach consists of identifying the properties of different neural processes, and from here inferring the properties of the phenomena they give rise to. This can be described as a bottom-up approach, which relies on the investigation of neural properties to suggest what kind of higher-level phenomena they produce.

After a brief outline of the mechanisms and target phenomena that GNWT and the Neural Stance provide, the mechanisms are assessed in terms of how well they are demarcated, and whether they in fact produce the phenomenon that their proponents suppose them to. It will be argued that both accounts fail. Instead, the valid application of standard demarcation criteria identifies quite different mechanisms, for quite different phenomena. The general problems that face any top-down or bottom-up approach to establish a mechanism for consciousness, and therefore how likely it is that consciousness can pick out scientific kinds, are then discussed.

6.4 Global Neuronal Workspace Theory

Dehaene and Naccache (2001; see also Dehaene et al. 2006) state that consciousness is equivalent to global availability, or the presence of information in a 'global workspace'. According to GNWT, long-distance 'workspace' neurons transmit attentionally amplified information from modular processing areas to multiple other areas. It is the breakdown of the usual modular processing that constitutes the global workspace. One important feature of global availability is that it makes information reportable, so reportability is taken to be a marker of global availability, and therefore of consciousness. The mechanism underlying global availability and reportability is suggested by Dehaene and Changeux (2004):

> Top-down attentional amplification is the mechanism by which modular processors can be temporarily mobilized and made available to the global workspace, and therefore enter into consciousness. (p. 1147)

> The model emphasizes the role of distributed neurons with long-distance connections, particularly dense in prefrontal, cingulate, and parietal regions, which are capable of interconnecting multiple specialized processors and can broadcast signals at the brain scale in a spontaneous and sudden manner... This broadcasting creates a global availability that, according to our hypothesis, is experienced as consciousness and results in reportability. (p. 1146)

The mechanism of global availability that results in consciousness and reportability consists of workspace neurons found in particular brain areas transmitting selected information across processing modules. Information is selected either through top-down or bottom-up attention and is broadcast for further processing and report.

Given the popularity of Global Workspace theories, it would appear that the mechanism of global availability is well demarcated. As summarised in Chap. 5, there is a range of evidence from attentional blink and backward masking paradigms that global availability, in the form of late ERPs such as the P300, widespread activity, and late, sustained and high-frequency neural synchrony correlate with conscious perception. However, as discussed earlier these correlations are dependent on the particular kind of masking, response type, and data analysis used in the small set of experimental protocols that GNWT uses to operationalise consciousness. In particular, supporters of GNWT argue that early stages of neural activity may form part of the total NCC or mechanism, but not the core NCC or mechanism of consciousness.

6.4.1 The Many Forms of Reportability

To support the argument that only late, global neural activity can be part of the (core) mechanism for consciousness, GNWT relies on a particular method of demarcating relevant neural correlates, and therefore relevant mechanistic components, from necessary background conditions. Sergent and Dehaene (2004a, 2004b; see also Del Cul et al. 2007; Sergent et al. 2005) found that in the attentional blink and backward masking paradigms, subjects' responses tended to be bimodal, favouring either 'clearly seen' or 'clearly absent' responses. This was found even though subjects were able to use a continuous scale of visibility ratings in other paradigms. Proponents of GNWT identify the mechanism for consciousness with whatever components also exhibit this bimodal distribution in activity.

However, a crucial problem with this method is that reportability itself does not pick out a single, stable, phenomenon, as different ways of gathering reports generate different distributions of responses. This is important because, as seen in the previous chapter, the different ways that responses are gathered have consequences for the kind of neurophysiological activity that is said to correlate with reports. As described in Chap. 5, those searching for neural correlates for a continuous distribution of responses will identify a much wider set of components than those identifying a mechanism based on the correlates for a bimodal response distribution.

For example, it is mainly late ERPs and neural synchrony in fronto-parietal regions that exhibit a bimodal distribution of activity for the attentional blink. However, earlier ERPs and early neural synchrony across other parts of the brain are also identified as relevant mechanistic activities and components when continuous response distributions are found in other experimental paradigms. Given this variation in how correlates and mechanisms can be identified, Deheane's group must offer justification for their particular method of gathering responses. That is, they must provide justification for the way they operationalise consciousness as a particular *type* of reportability (i.e., all-or-nothing responses related to the attentional blink).

Yet as suggested in Chap. 2 there is no 'best' method of gathering subjective responses, just different ways that exhibit the different effects of response criteria. The response biases that generate different response distributions cannot be eradicated, just altered or maximised towards reaching a particular goal (e.g. fewer false alarms). Further, the factors that determine reportability vary across paradigms, including the factors of different task instructions, response scales, training, expectations, and so on. Melloni et al. (2007) note that conflicting claims about the relevant components for a mechanism of consciousness can be partly explained as a consequence of the different types of tasks used in experimental paradigms, such as those that rely on attention and those that do not.

The variation in the correlates found for reportability across a range of experimental paradigms, combined with the fact that there is no 'best' way of gathering reports, suggests that 'reportability' does not in fact refer to a single phenomenon, and is not realised by a single mechanism. The generation of reports given in particular tasks can draw differently on the components serving depth and complexity of processing, inhibitory processes, working memory, spatial or object-based attention, and decision-making, and is sensitive to a range of different factors, such as motivation, task instructions, and so on. This means that different neural components should be considered as relevant components of a mechanism of reportability, depending on which sensory or cognitive processes, working in which context, are under investigation. 'Reportability' does not refer to a single phenomenon, but to a range of task-specific phenomena, produced by a range of task-specific mechanisms.

This means that it is not the case that researchers investigating the correlates of reportability are simply labelling different sections of the same mechanism as 'core' or 'background' parts of the mechanism of consciousness. There is not, as it is often supposed, a significant problem in identifying necessary and sufficient conditions for reportability, or total vs. core correlates, or background and constitutive mechanistic components for reportability. Researchers arguing that early neural activity forms part of the mechanism of reportability use a perfectly valid methodology to identify the mechanism for a very specific phenomenon; the type of response investigated in a particular experimental paradigm. It is not surprising that different neural correlates are found for different parts (e.g. pre/post attentional) of different perceptual processes (e.g. detection vs. match to sample tasks) assessed in different ways (e.g. by using objective/subjective measures). The disagreement is not over identifying 'background' or 'core' components for the same phenomenon, but over different phenomena altogether.

The specific way that GNWT operationalises consciousness and the way it subsequently establishes the mechanism for reportability is not the only way to investigate the production of reports of visual stimuli. Neither can it be argued that equating the phenomenon of 'reportability' with bimodal response distributions in attentionally demanding tasks is the 'best' way. There are other ways of investigating the conditions under which reports of different types, in different tasks, are generated. That 'reportability' is itself a pluralistic phenomenon, and is not realised by a single mechanism, forms the first major problem with GNWT. A second problem with GNWT is discussed below, where it is argued that Dehaene and colleagues misuse standard demarcation criteria that instead identify quite different mechanisms for different phenomena.

6.4.2 Demarcating Mechanisms of Reportability

An important feature of GNWT is the way in which it demarcates the mechanism of consciousness (reportability) from background conditions. Early instances of global availability are clearly necessary for reportability (whatever type), but according to GNWT they should not be considered as a constitutive part of the mechanism for consciousness, but as a background condition. However, as shown in the previous chapter, and above, the way that the correlates and mechanisms of consciousness are identified depends on what type of reportability is used to operationalise it. This section extends the arguments above by considering how Craver's demarcation criteria apply to a range of types of reportability. It will be argued that when standard demarcation criteria are applied, very different mechanisms of reportability are found than that proposed in GNWT. While GNWT may successfully identify part of a mechanism that produces bimodal response distributions in attentional blink paradigms, well-demarcated mechanisms for other report-related phenomena are much more inclusive.

Craver's criteria for constitutive relevance, based on mutual manipulability, offer a structured way of assessing the claim that the mechanisms for reportability, whatever their type, do not include early activity from parts of the visual system, or early instances of neural synchrony. As stated earlier, the criteria of constitutive relevance for a component are that stimulating the component stimulates the phenomena in question, that changes in the phenomena can be tracked by changes in the component, and that these mutual manipulations are both specific and subtle. These criteria are applied below to assess whether a well-demarcated mechanism for reportability can include early neural activity.

It is crucial to note that the following analysis proceeds by largely ignoring theoretical preconceptions about what the mechanisms of consciousness *should* look like, and instead focuses on how standard demarcation criteria apply to real experimental practices and results. This means that if there are suitably specific and subtle mutual manipulations between a candidate component and a type of reportability, then it will be identified as a component in the mechanism for this type of reportability, regardless of how this fits in with theoretical assumptions about consciousness. This analysis will also be used to further criticise the validity of the distinction between the correlates/mechanisms of the contents of consciousness, and correlates/mechanisms of consciousness of those contents, outlined in earlier sections.

The first criterion for something to count as a mechanistic component for a given phenomenon is that, when altered it alters the target phenomenon, and when stimulated, stimulates the target phenomenon. As accepted by all researchers, interfering sufficiently with early neural activity will affect whether or not a stimulus is reported, as it is clear that this early activity at least forms a necessary component in the entire process. For example, knocking out V1 produces 'cortical blindness' and ensures that no (or at least very few) visual stimuli are reported. It also appears that more localised lesions to early visual areas, including temporary lesions delivered by TMS, also prevent subjects from reporting specific stimuli (e.g. Koivisto et al. 2011 briefly discussed in Chap. 5). This means that the first part of the first criterion is fulfilled; changes in early activity result in changes in responses.

It is also true that stimulation of early neural activity can stimulate later stages of neural activity that results in reportability. This suggests that certain types of early neural activity may form relevant components of the mechanism for reportability.

This suggestion can actually be found in GNWT, in which strong bottom-up signals are sometimes sufficient to 'capture' top-down attention and thereby enter the global workspace. In this case, stimulation of relevant types of early neural activity is sometimes sufficient to stimulate consciousness and report.

In fact, stimulation experiments are an important way of investigating the functional structure of visual cortex. Much of the work using cortical electrical stimulation is still carried out on non-human primates (e.g. Britten and Wezel 1998; Afraz et al. 2006), where the effects of stimulation on judgments about stimuli (e.g. direction of motion, face categorisation) are measured. Using the same techniques on epileptic human patients offers the opportunity of both measuring the threshold above which forced-choice detection occurs, but also in investigating the contents of subjects' reports about visual information following localised electrical stimulation. By using reports to assess perceptual processes, these paradigms offer an alternative way of assessing the claim that stimulation of early neural areas does or does not stimulate reports of stimuli.

For example, Lee et al. (2000) stimulated areas across occipital cortex and categorized subjects' reports as being about simple, intermediate or complex forms, coloured forms, and motion. In this way they were able to map out the human equivalents of visual shape, colour and motion areas already found in macaque monkeys. More recently, Murphey et al. (2009) used fMRI to identify different functional areas and investigated how subjects would respond to electrical stimulation in those areas. They found that stimulation of early visual areas (V1–V3) almost always resulted in successful detection, as well as producing a report of a visual sensation, but that stimulation of later areas typically did not. That is, stimulation of early visual areas stimulated the production of reports. They suggest that "there is no sharp dichotomy between early and late visual areas in their ability to support perception" but that stimulation of early visual areas may more often lead to reports "because of greater extrinsic connectivity in early areas", thus allowing propagation across the perceptual network more easily (Murphey et al. 2009, p. 5391).

This finding has several implications. First, stimulation studies of early visual areas show that stimulating early neural activity is sufficient to stimulate a chain of processes that result in reports: stimulation of areas V1–V3 almost always stimulates reports. According to the stimulation condition above, a component that stimulates the target phenomenon when it is stimulated may form part of the mechanism for that component. In this case, areas V1–V3 fulfill the first criterion for being a constitutive component in the mechanism that generates reports about various types of visual stimuli.

Second, the finding that stimulation in later visual areas *only* does not reliably stimulate reports suggests that activity in early visual areas is necessary to promote the production of reports. The greater connectivity of early visual areas means that early neural activity plays an essential role in spreading information across the brain; global availability and reportability is not something that is accomplished solely via long-distance 'workspace' neurons. The kinds of report generation investigated in these experiments cannot occur without the participation of early visual areas, and in a way that suggests again that they are constitutive components, not merely components fulfilling background conditions. Murphey et al. also suggest that there is little difference in terms of 'supporting' perception, measured using reports, between early and late visual areas. This idea that early and late neural activity are not easily decomposable into background and constituent components for a mechanism that generates reports of visual stimuli is discussed further below in relation to Lamme's Neural Stance. For now it is sufficient to note that early neural activity satisfies the stimulation condition for a constitutive component for mechanisms of the reportability of visual stimuli.

The second criterion for constitutive relevance is that changes in the phenomenon under investigation change the component in question. This criterion can most easily be tested across instances of 'seen' and 'unseen' trials by comparing levels of early neural activity. As noted in more detail in Chap. 5, Koivisto and Revonsuo (2003; see also Koivisto et al. 2005, 2006) found that an early ERP (200 ms after stimulus onset) differentiated 'seen' from 'unseen' trials. Melloni et al. (2007) found that reported stimuli, but not unreported stimuli, generated early, global, higher power and phase synchrony gamma band oscillations. Palva et al. (2005) also found that early, strong, global activity was only found for reported stimuli. Dehaene and colleagues also find these differences in early neural activity across 'seen' and 'unseen' trials, but use their specific method of searching for bimodal distributions of activity in a particular paradigm to identify the 'core' correlates of consciousness. In either case, changes in whether reports are generated are clearly reflected in the differences between the occurrence of certain types of early ERPs, the relative strength and phase locking of early global synchrony, and activity in early visual areas.

Cases of binocular rivalry offer other interesting test cases in which different stimuli are shown to different eyes, resulting in shifts in which stimulus is reported. This allows stimuli to be kept identical with only a shift in what is reported. Logothetis and Schall (1989) famously showed that during binocular rivalry, cells in V5/MT in monkeys were the earliest cells to show a difference in activity between 'seen' and 'unseen' stimuli. Moutoussis et al. (2005) found that even earlier cells in humans showed a difference in activity for 'seen' (reported) and 'unseen' (unreported) trials (down to V3A and LOC). They suggest that perception of particular stimuli (in this case moving stimuli) should be attributed to a distributed set of brain areas and over both early and late brain activity. Therefore both 'normal' cases and binocular rivalry show that early local neural activity tracks whether or not visual stimuli are reported. These early components therefore fulfill the second criterion for inclusion into the mechanism for the generation of reports.

Two other criteria for demarcating constitutively relevant components of a mechanism from its background conditions are that changes in the component produce both specific and subtle changes in the phenomenon that the mechanism is supposed to produce. Just how specific and subtle these changes need to be is debatable, but the evidence presented so far (including that from previous chapters) suggests that the relationship between certain sorts of early neural activity and reportability satisfies these constraints. In general terms, strong early neural activity (including pre-stimulus activity) is likely to affect the occurrence of later instances of global synchrony. Melloni et al. (2007) state this early activity "may be a correlate of the anticipation of the matching between short-term memory contents and sensory input" (p. 2864). That is, early synchrony may instantiate expectations that bias subsequent processing. Synchrony that biases subsequent processing will not only affect *if* a stimulus is reported by directing attentional allocation, but also *what* kind of stimulus is reported. Palva et al. (2005) suggest that the probability of a report being generated, related to a specific stimulus, can be affected by a wide range of temporal factors, including:

> . . . fatigue and changes in arousal, by fluctuations in attention, by variable accuracy of selective attention and of short-term memories of preceding stimuli, by intermittent prestimulus cortical states, and finally, by various factors in poststimulus neural processing ranging from early top-down modulation to uncertainty in decision making. (Palva et al. 2005, p. 5255)

In relation to more specific cases of report generation, such as the example of motion detection discussed earlier, early neural activity clearly forms part of the mechanism of reportability of that stimulus. Activity in V5/MT is related in a specific and subtle way with reports of motion, and stimulus-specific activity in other early visual areas, as suggested above, are also related with specific reports in this way. Whether or not a stimulus is reported is affected by just what sort of task and stimuli are involved. In rejecting the idea that 'reportability' refers to a single phenomenon, it should not be surprising that the task-specific mechanisms for responding to different stimuli in different ways are likely to include early, task-specific components.

Early neural activity forms a constitutive component of many task-specific mechanisms for particular instances of report generation. Stimulation experiments involving report are routinely used to investigate the functional structure of visual cortex, and show that stimulation of early neural activity is sufficient to drive the processes that generate task-relevant reports. Changes in different types of reportability are also tracked by changes in early neural activity in particular brain areas and through different types of global neural synchrony. Finally, changes in early neural activity can produce specific and subtle changes in what subjects report.

As the mechanisms for perceiving a stimulus, indexed by report, include these instances of early neural activity, it is hard to see why mechanisms of consciousness, also indexed by report, should leave these components out as mere background conditions. Once it is seen that report generation is a task and stimulus specific phenomenon, this move is simply not consistent with empirical findings. That is, the claim that other proposed mechanisms of report generation merely identify background conditions for consciousness is based on a misidentified target phenomenon—reportability *per se*. Other proposed mechanisms are mechanisms for the generation of other kinds of task-specific reports, and the mechanisms of report generation are far broader and more inclusive than supposed by GNWT.

6.4.3 Contents and Consciousness

One obvious retort to this argument is to invoke the distinction between the correlates/mechanisms of the *contents* of consciousness, and the correlates/mechanisms of *consciousness of that content.* The critic would argue that all the stimulations and changes in early neural activity that are tracked in reports are simply changes in the contents of consciousness, but are not part of the mechanism that produces consciousness of those contents. For example it would be claimed that stimulation of early visual areas makes it more likely that certain sorts of content become conscious, but the mechanism for making these contents become conscious consists of later neural activity across fronto-parietal areas. Likewise, changes in early neural activity that track changes in report generation are simply reflecting changes in the contents of reports, but not changes in consciousness of that content, which is determined by much later instances of neural activity. Therefore it could be argued that the mechanisms for processing content and the mechanisms that produce consciousness of that content are different, and divided between early and late neural activity.

One simple response to this claim is to ask for some evidence that there is indeed a valid distinction to be drawn between those mechanisms that generate content, and those that generate consciousness (or reports) of that content. That is, if the mechanisms that produce content and the mechanisms that produce consciousness of that content can be distinguished experimentally, then there are plausibly two different sets of mechanisms. This however may be rather difficult.

As noted above, not only changes in the contents of reports, but changes in whether content is reported or not, can be tracked and stimulated in terms of early neural activity. For example, stimulations of early processing areas (e.g. V1–V3) stimulate, in a fairly specific and subtle way, the *contents* of reports. Also, *whether or not* particular contents are reported can be tracked, again in a reasonably specific and subtle way, through changes in early neural activity. This is entirely to be expected in a complex highly interconnected system that relies on recurrent processing and neural synchrony to function. Early processing does not merely provide content that is then taken away for further processing in other areas. Repeating Murphey et al. (2009) from above: "there is no sharp dichotomy between early and late visual areas in their ability to support perception" (p. 5391).

However, the problem may be more a conceptual one than an empirical one. Clearly something is going on in attempts to separate neural activity into earlier 'content' stages and later 'consciousness of content' stages of processing. There are stages of neural activity that are more subtly and specifically related to stimulus specific processing, or to the contents and distribution of subjects' reports. Indeed, it is possible to loosely decompose the stages of processing present in particular instances of report generation. For example, the mechanism of report generation for a particular visual stimuli can be decomposed into specific types of sensory processing, attentional allocation, decision making, motor planning, and so on. Manipulations of response criteria can be used to isolate activity that is more or less related to stimuli-specific processing, and that activity related to decision-making. Importantly, this decomposition uses terms that directly refer to the stages necessary to process information and decide how to act on it, yet acknowledges that these processes may have no sharp boundaries.

Hulme et al.'s (2008) work on dissociating the correlates of 'reportability' briefly noted in the previous chapter, is a good example of this approach. Without making assumptions about the relationship between stimulus reportability and mechanisms of consciousness, they decomposed the neural activity that led to reports of stimuli into the stages of stimulus processing, decision-making and motor report. These decompositions are made using experimental manipulations and the application of Signal Detection Theory to dissociate different processes. However, to label parts of this framework as mechanisms for processing content, and mechanisms for processing consciousness of that content, is misleading. They are simply different stages of the process of report generation, and the components that fulfill these different roles are likely to be task-specific. As Hulme et al. (2008) state:

> Our results... identify networks of areas whose activations should be interpreted as reflecting components of report rather than as general markers of consciousness... performing this dissociation depends critically on the type of task, which in turn can determine whether performance is limited by early or late noise. (p. 1609)

That is, taking reports of stimuli as markers of consciousness does not allow a decomposition of processing into stages, and the localization of neural mechanisms, for processing content, and consciousness of that content. There is no single switch-point at which *any* content becomes 'reportable' (conscious) or not; the entire process, taken as a whole, determines what reports are generated. The only methodologically valid way to decompose mechanisms that generate stimulus-specific reports in specific paradigms, is into these stages of sensory processing, decision-making and motor report. None of these are obviously (or at least non-controversially) relevant to consciousness.

Further, the different components that fulfill these roles across different tasks make it difficult to establish any context-independent mechanism for generating consciousness of content. So, even if decision-making were taken to equate with consciousness, there is no single mechanism of perceptual decision-making, but different local mechanisms for different sorts of stimuli and tasks. The continued use of the distinction between content, and consciousness of content, only seems to result from the assumption that this distinction *must* be coherent at the physical level.

This provides further support for the claim that GNWT fails to identify a mechanism that can be equated with consciousness. First, reportability is not a viable target phenomenon by which to operationalise consciousness as a scientific kind. 'Reportability' itself refers to a wide range of report-related phenomena, and a wide range of task-specific mechanisms. Further, the distinction between content NCCs/mechanisms and NCCs/mechanisms for consciousness of that content is parasitic on, yet ignores, the methodologically valid decompositions of processing related to stimulus reportability, as described in contemporary cognitive science. The kinds of consciousness science are therefore not kinds of consciousness, but a wide range of well-disguised kinds smuggled in from cognitive science, including attentional phenomena, stimulus-specific sensory processing, and task-specific decision-making. Finally, the question of whether reportability is even a relevant target phenomenon for consciousness science is discussed below.

6.4.4 Identifying an Appropriate Function

All mechanisms have a function: "Mechanisms are identified and individuated by the activities and entities that constitute them, by their start and finish conditions, and by their functional roles" (Machamer et al. 2000, p. 6). Part of the demarcation conditions of a mechanism for consciousness must therefore make reference to the function of the mechanism (i.e. whether the components together fulfill this function), and it must be established that the function of this mechanism is appropriate for a mechanism of consciousness. As seen in this and subsequent sections, this criterion for a mechanism of consciousness threatens both Dehaene's and Lamme's accounts.

Although not couched in explicitly functionalist terms, consciousness in GNWT is equivalent to the function of making information available to a wide range of cognitive processes, and the identification of consciousness *as* reportability is necessary for its operationalisation under GNWT. In claiming to have found the mechanism for reportability, Dehaene and colleagues can therefore claim that they have also found the mechanism for consciousness. However, Lamme (2006) argues that such functional identifications of consciousness miss the point. All that can be learned from these functional descriptions and mechanisms of consciousness are details about these specific cognitive functions (e.g. attention, working memory, etc).

Lamme, following Block, argues that the having of experience, or phenomenal consciousness, should not be identified with any cognitive function at all. Instead, he states that phenomenal consciousness is associated with an early, pre-attentional stage of processing that is later 'accessed' and made available for report via the sort of cognitive functions that GNWT identifies as consciousness. Thus, Lamme argues that GNWT and the mechanism it offers is not a mechanism for consciousness at all, but a mechanism for the cognitive functions that are used to 'access' experience, such as attention and working memory. In virtue of the way consciousness is functionalised and operationalised in GNWT, the measures and mechanisms it offers are therefore pitched at the wrong target.

However, supporters of GNWT offer some pragmatic responses to Lamme's objection. Dehaene et al. (2006), in discussing the possibility of unreported or unattended contents of experience, state: "Whether [subjects] actually had a conscious phenomenal experience but no possibility of reporting it, does not seem to be, at this stage, a scientifically addressable question". (p. 209; see also Dehaene and Changeux 2004). Kouider et al. (2007) echo this methodological worry: "Given the lack of a scientific criterion, at this stage at least, for defining conscious processing without reportability, the dissociation between access and phenomenal consciousness remains largely speculative and even possibly immune to scientific investigation" (p. 2028). Thus, while Lamme may raise an interesting objection, Dehaene and his colleagues argue that the method of ascribing a function to consciousness and operationalising it via some sort of behaviour is necessary to do *any* science of consciousness. In this case, using the working definition of consciousness as reportability is the only way forward.

It will be argued in later sections that the problem of providing an adequate functional description of consciousness, or some other way of identifying it, points to a very different conclusion to either of those offered by Lamme or Dehaene. However, it is first necessary to assess Lamme's proposed mechanism of phenomenal consciousness based on his 'Neural Stance'. He argues that by taking the distinction between two types of neural processing seriously, they can be used to support the distinction between access and phenomenal consciousness, independently of the presence or absence of reports. While Lamme's Neural Stance is more in line with a mechanistic approach, and although it offers to solve some of the problems in establishing empirically supported demarcations and distinctions found in GNWT, it fails for very similar reasons.

6.5 The Neural Stance: Local Recurrent Processing

Lamme's Neural Stance (2004, 2006) uses the distinction between different kinds of neural information processing to demarcate the mechanisms for particular cognitive functions (e.g. working memory, attention) from the mechanism of phenomenal consciousness. He argues that the differences between feedforward and recurrent processing present a scientific way of demarcating conscious (recurrent) from unconscious (feedforward) processing. Lamme argues further that it is recurrent processing itself, not recurrent processing found in specific locations (fronto-parietal network) at specific times (later stages of processing) and related to specific cognitive functions (e.g. attention or report) that provides the mechanism for consciousness. This bottom-up method of identifying neuroscientific categories of phenomena and then matching them with higher level phenomena is a promising alternative method for demarcating a mechanism for consciousness. It comes with the methodological advantage that it ignores the strong ties with problematic behavioural operationalisations of consciousness discussed in earlier chapters.

According to Lamme's model, populations of neurons engaging in recurrent processing of sufficient strength, and thereby triggering synaptic plasticity processes including learning and memory, form the mechanism for consciousness. The focus on the activity of recurrent processing, wherever its location, is a much simpler way of isolating the mechanism of consciousness than the GNWT account above, and its simplicity is appealing:

> We could even define consciousness as recurrent processing. This could shed an entirely different light on the matter of whether there is conscious phenomenal experience in states of inattention, split brain or extinction. The matter would now become a scientific debate, where evidence from behavioral observations is weighed against the evidence from neural measurements. If recurrent interactions of sufficient strength are demonstrated, it can be argued that the 'inattentional', 'preconscious' or 'not reported' still have the key neural signatures of what would otherwise be called conscious processing. (Lamme 2006, p. 499)

However, Lamme places further constraints on what sort of recurrent processing counts as conscious processing, which opens up his account to criticism. Lamme argues that whenever neural populations exhibit recurrent processing (abbreviated to RP throughout), consciousness ensues, no matter where in the brain RP occurs. However, he then states that it is only early *local* cases of RP that are the mechanism for (phenomenal) consciousness, with later global cases of RP correlating with attention and other processes related to reportability, or access to consciousness. Therefore the claim that phenomenal consciousness *is* recurrent processing is imprecise; it is only early instances of local RP that Lamme argues form the mechanism for phenomenal consciousness. Whether local recurrent processing is in fact a sufficiently well-demarcated mechanism, and whether it provides a mechanism for phenomenal consciousness, is discussed below.

6.5.1 Local vs. Global Recurrent Processing

Craver's criteria of mutual manipulability cannot be applied to Lamme's identification of local RP with phenomenal consciousness in the same way as it was applied to GNWT above. This is because Lamme's approach, as discussed in more detail in the previous chapter, is a decidedly bottom-up approach. Instead of specifying a behavioural operationalisation of consciousness, and searching for the mechanism for this target phenomenon, Lamme's approach is to identify a type of neural processing, and use the properties of this process to suggest the phenomenon that it gives rise to (phenomenal consciousness). Without a specification of the target phenomenon, and in letting the neural mechanism determine what the target phenomenon is, it does not make sense to apply criteria of mutual manipulability to local RP and phenomenal consciousness. That is, if it is stipulated that local RP is the mechanism for phenomenal consciousness, and the properties of consciousness are determined entirely by the properties of local RP, then criteria of mutual manipulability are satisfied by definition. There could be no possible mis-matches between stimulations of local RP and changes in phenomenal consciousness because consciousness just is whatever local RP produces. In this case, evaluating the claim that local RP is the mechanism for phenomenal consciousness must be done on different grounds. Discussed further below, these include different ways of questioning the internal coherence of Lamme's claims.

Lamme initially argues that it is the mere presence of RP that is important to consciousness. RP enables synaptic plasticity and learning, while feedforward processing does not. The distinction between RP and feedforward processing therefore sounds like a candidate distinction between conscious and unconscious processing. Further, if RP is all that matters in ascribing instances of consciousness, then whether it is present in early and local or late and global stages of processing should make no difference. However, Lamme also argues that it is *only* early and local RP that is relevant to phenomenal consciousness, while global RP gives rise to access to consciousness. Isolating only early stages of RP as the mechanism for phenomenal consciousness directly contradicts the idea that it is RP itself, and the way it triggers synaptic plasticity, that is the mechanism for phenomenal consciousness. In fact, isolating local RP as the mechanism for consciousness directly contradicts the argument Lamme invokes to identify RP with consciousness in the first place:

> ... stimuli that evoke RP change your brain, while stimuli that evoke only feedforward activation have no lasting impact. It would be difficult to see why the involvement of the frontoparietal network would make such a difference (after all, it is all neurons firing action potentials, whether they are in the back or front of the head). (Lamme 2006, p. 499)

If 'it is all neurons firing action potentials', then neurons engaging in recurrent processing in *both* the front and the back of the head should be identifiable with phenomenal consciousness. That is, if recurrent processing is what matters then recurrent processing, whether limited to local sensory areas *or* extended across fronto-parietal regions, will both result in phenomenal consciousness. By identifying phenomenal consciousness with a mechanistic *activity* (recurrent processing), and rejecting the method of associating consciousness with specific *entities* or brain areas (e.g. neurons in fronto-parietal network), Lamme has no grounds on which to restrict the relevant performance of this activity for phenomenal consciousness to local areas only.

Also, as noted above, early global synchrony (i.e. early global RP) may actually play an essential role in driving and sustaining early local recurrent activity. Anticipatory global RP biases and directs a system's subsequent processing, so the likelihood and 'content' of local RP is therefore partly determined by early (including pre-stimulus) instances of global RP. The assumption that local RP occurs first, delivering phenomenal consciousness, and is then followed by global RP, which delivers access to consciousness, as found in Lamme's model of cortical processing, is therefore a problematic one. Without a clean dissociation between early local RP, and late global RP, the distinction between an early, local pre-attentional stage of phenomenal consciousness and a later, global attention-based stage of access to consciousness is severely compromised.

Another potential problem with Lamme's account is discussed below, where it is questioned whether the implicit assumptions found in his descriptions of conscious content, and the contents that local RP actually provides, are consistent with each other.

6.5.2 *Questions of Content*

One possible problem for Lamme's account is that local RP is an incredibly widespread mechanism, and as such runs the risk that it provides a very counter-intuitive range of conscious content. While the Neural Stance apparently does away with intuitions about consciousness, the content picked out by local RP conflicts with the way in which Lamme himself describes the contents of phenomenal consciousness. This at least suggests that the real implications of taking the Neural Stance have not yet been fully elaborated. Currently, it is unclear if these implications are such that they could still be stomached by Lamme, and therefore if local RP should be identified with phenomenal consciousness. While not conclusive, this section argues that there is more work to be done in understanding Lamme's claims.

Lamme sees the fact that local RP picks out a wide range of content as a positive aspect of taking the Neural Stance. Local RP occurs from very early on in stimulus-specific processing, and can be found in many brain areas and many stages of processing concurrently. This could potentially explain why we think we see much more than we can report, as it allows an identification of more conscious content than found in the global workspace (i.e. what is attended and reported). In reference to an inattentional blindness paradigm that uses a stream of letters to 'blind' people to textured squares surrounding the fixation point (Scholte et al. 2006), Lamme states:

> Importantly, the proposal allows for multiple complexes to exist at the same time. In the IB [inattentional blindness] experiment one complex could therefore represent the attended stream of letters, while another would represent the not-reported objects in the background. By definition, both would be conscious representations. (Lamme 2006, p. 499)

Local RP as the mechanism for phenomenal consciousness therefore allows findings from change and inattentional blindness paradigms to be explained using a two-tier model. The stimuli that subjects report are those few that receive top-down attention and are maintained in working memory. However, these stimuli are only a subset of the contents of phenomenal consciousness, which also includes contents from all stimulus information that is subject to local RP. The discrepancy between the amount of content picked out by local RP (and is therefore phenomenally conscious) and the amount of content that can be sustained in working memory at any one time explains why the contents of our experiences seem to 'overflow' our reports of them.

However, while the identification of local RP as the mechanism for phenomenal consciousness may explain why we report seeing more than we can actually identify, local RP would in fact provide far more phenomenal content than Lamme appears to endorse. Even though he argues that the Neural Stance should determine how the contents of consciousness are identified, the content picked out by local RP is vast. RP has been found in V1 for surface segregation (Scholte et al. 2008), figure-ground perception (Supèr et al. 2003), and orientation (Boehler et al. 2008), suggesting that the contents of consciousness (somehow) include some of the earliest stages of visual processing.

Further, models of perception as an inferential process (e.g. Lee 2002; Mumford 1992) show that RP plays a crucial role in mediating differences between expectations and actual input, and in resolving competitions between different interpretations of the same input. Recurrent processing serves to minimise these differences across and between hierarchies of neural processing in order to establish a 'most likely' hypothesis that makes sense of existing expectations and the actual input. The role of local RP is precisely to resolve competitions between different interpretations of input. Therefore, if local RP is the mechanism for consciousness, then Lamme is committed to the claim that subjects will be conscious of all the competing information carried in these instances of RP, and all at the same time.

This multiplies the contents of phenomenal consciousness far beyond the content Lamme attributes to subjects in the inattentional blindness paradigm. In the quotation above he claims that one complex of local RP can represent attended letters, and another complex represent unreported items in the background. In describing the contents of phenomenal consciousness, Lamme uses the kind of language that preserves some common phenomenological intuitions about conscious content. These include the idea that there is at most one representation of an object conscious at any one time, and that it is a reasonably detailed and (probably) object-level representation.

Yet if local RP is the mechanism for phenomenal consciousness, and local RP can be found for all neurally instantiated interpretations of input (which seems to be the case), there will not be a single representation of either the attended letters, or the unreported items. Instead, there will be a number of more or less active, and inconsistent, representations of the same set of visual information, at different levels of processing (e.g. shape, colour, identity), all in existence at the same time. The existence of multiple sets of overlapping and inconsistent phenomenal representations of visual information is however not a possibility that is entertained in Lamme's discussions. There is instead an implicit assumption that for each object that evokes local RP, there is a single 'representation' of the object that is phenomenally conscious. The function of local RP suggests that this is far from the case.

The range of different overlapping, inconsistent and multi-level content provided by local RP is difficult to interpret in phenomenal terms. This does not of course conclusively show that local RP should not be identified as the mechanism for consciousness. If we are to truly take the Neural Stance, then it will be necessary to bite the bullet and agree that all instances of local RP, in whatever form, contribute to the contents of phenomenal consciousness. Perhaps these contents are not experienced in such a way that they can be reflected on or reported, which would explain why they appear so counterintuitive.

Yet Lamme does not explicitly acknowledge anywhere that the real range of content picked out by local RP is a plausible range of content for consciousness. In retaining somewhat standard phenomenological language, and claiming that the visual system delivers single phenomenal representations of objects, Lamme appears to endorse the same conception of phenomenal consciousness that many others do, i.e. that of providing single, high-level representations of objects. The real extent of the contents of local RP is not something that can be accommodated within this language (this theme is developed further in Chap. 7 and 8).

Therefore identifying local RP with phenomenal consciousness requires a much deeper conceptual shift than suggested by Lamme. Taking local RP seriously is not consistent with the idea that representations of unattended and unreported objects are present in a detailed, picture-like phenomenal consciousness. Instead, it entails that multiple and competing representations of the same content across all levels of processing are experienced at the same time. If Lamme can accept this range of conscious content, and the real conceptual shift in talking about consciousness that it requires, then the position is at least consistent. Currently though, it is not clear if the real implications of the Neural Stance have been acknowledged. This leaves the position underdeveloped at best, and potentially seriously problematic. Just what local RP is the mechanism for is the subject of the next section.

6.5.3 The Function of Recurrent Processing: Running the Argument Both Ways

One of the main motivations behind Lamme's identification of local RP as the mechanism for consciousness is to avoid claiming that consciousness is identical with the cognitive functions by which it is operationalised. In this way, Lamme hopes to establish a mechanism for Block's phenomenal consciousness rather than ways of 'accessing' consciousness via memory, behaviours, and reports. As Lamme cannot rely on a direct operationalisation of phenomenal consciousness in order to identify its mechanism, he relies instead on an inference to the best explanation. Both local and global RP are always present during clear-cut episodes of consciousness (i.e. report), but late global availability is argued to be the mechanism for the confounding factors of attention and working memory. Once these confounding factors are taken away, only local RP is left, so local RP is the mechanism for phenomenal consciousness. Therefore, Lamme concludes that the presence of local RP always determines the presence of consciousness, even when a subject cannot report the content of the experience.

However, local RP cannot be described as a non-functional element either in the cognitive processes that enable reportability, or in cortical information processing more generally. As noted above, local RP makes use of feedback connections that enable comparisons to be made between expectations and input between different levels of neural processing, and between different interpretations of the same input. Through this, a 'winning interpretation' is produced which goes on to be used in later processing. Accordingly, local RP plays a very clear functional role in the selection of appropriate interpretations of (sensory) input. Given this, it may be possible to run a similar argument against Lamme's mechanism of local RP to that used against GNWT:

P_1) Global availability (late RP) can be described in terms of the specific cognitive functions used in operationalisations of consciousness.
P_2) Consciousness is not equivalent to the cognitive functions used to operationalise it.
→ C_1) Global availability is not the mechanism for consciousness.

P_3) Local RP can be described in terms of its function in information processing, i.e. to enable 'most likely' interpretations of input to emerge from neural processing.
P_4) Consciousness is not equivalent to the function of enabling 'most likely' interpretations of input to emerge from neural processing.
→ C_2) Local RP is not the mechanism for consciousness.

The functional descriptions offered for global availability and local RP are at different levels of description (cognitive abilities vs. neural processing), but it is difficult for either to escape the criticism that the mechanisms they describe produce something other than consciousness. The problems associated with functionalising consciousness are perhaps well-known, but from earlier chapters it should be clear why these problems arise.

In terms of the top-down approach to investigating mechanisms of consciousness, there is no experimentally manipulable function that can be assigned to consciousness that cannot be accounted for by (a set of) neural or cognitive mechanisms. There is no unique function that consciousness can fulfill that cannot be done by reference to mechanisms of learning, perceptual processing, decision-making, information integration, monitoring mechanisms, and so on. In terms of the bottom-up approach, well-demarcated neural mechanisms will also have a function, describable in terms that do not include consciousness. This strongly suggests that whatever mechanisms and scientific kinds are found in consciousness science are just those mechanisms and kinds that are found in the cognitive sciences, and not mechanisms of consciousness at all. This claim is elaborated on below.

6.6 Mechanisms and Scientific Kinds

The mechanisms described above suffer from many of the same problems. Both GNWT and the Neural Stance fail to generate well-demarcated mechanisms, and they both fail to accurately describe the phenomena that the mechanisms produce. The mechanisms that can be identified, using GNWT and the Neural Stance as starting points, are for quite different phenomena. Although this chapter has focused on only these two accounts, it is suggested that other attempts to locate a mechanism of consciousness will fail for similar reasons. In this case, consciousness science will fail to identify any scientific kinds, leaving 'consciousness' as a problematic and non-viable scientific term.

First, GNWT fails as a satisfactory mechanistic account of consciousness because the phenomenon that it uses to operationalise consciousness—reportability—is itself a pluralistic phenomenon. Reportability refers to many different phenomena, and mechanisms for different types of report generation are task-specific and are much more inclusive than supposed in GNWT. Across the various mechanisms that generate reports, the application of Craver's criteria of mutual manipulability shows that early neural activity constitutes a component in the many mechanisms of reportability, so is not merely a background condition. Further, the distinction between the correlates/mechanisms for the contents of consciousness, and correlates/mechanisms for consciousness of that content fails as an empirically sustainable distinction. Instead, the stages of stimulus-specific processing, task-specific decision-making, and response-specific motor planning together produce reports, but are made up of different components across different tasks. While reportability may seem like a reasonably straightforward way of operationalising consciousness, and one that picks out a single phenomenon and mechanism, this is not supported by the evidence reviewed above.

This means that consciousness, operationalised as reportability, does not pick out a scientific kind. GNWT can be described as a well-disguised research program about the effects of attention on global neural processing and report generation in attentional blink and inattentional blindness paradigms, but not as a research program about reportability *per se*. By their own definition, GNWT does not therefore constitute a research program about consciousness.

A possible response to this is that mechanisms of different types of reportability clearly have something in common in terms their decomposition at the *computational* level, (stimulus-specific processing, decision-making, and so on). However, the variety of ways that mechanisms of reportability are actually realized ensure that they fail to form a coherent scientific kind. Kinds allow reliable predictions and broad generalisations to be made about the phenomena they explain. However, the range of mechanisms for different types of report-related phenomena are sensitive to a wide range of task-specific factors, making it difficult to make predictions and generalisations about the occurrence and contents of reports. For example, the ways in which expectations affect subsequent processing and reports, the role of multiple types of exogenous and endogenous attention, motivational states of the subject, specific task demands, stimuli and cognitive competencies, the role of task design and response biases, and what indeed is taken to be a 'report' (verbal, button-press, forced-choice), all affect the probability of a report being in different situations.

Indeed, it is precisely *because* it is difficult to make reliable predictions and generalisations about subjects' performance at a range of perceptual and cognitive tasks, targeted at different types of stimuli, and indexed by different types of responses, that more specific competencies are the focus of cognitive science. Before consciousness science, there was no research field dedicated to the investigation of *the* mechanism of report generation, because report generation refers to so many very different phenomena. Seen in this light, attempting to locate one neural mechanism that determines how we respond to external stimuli *in general*, may seem like an odd research aim to have.

Second, Lamme's Neural Stance, while in itself an interesting theoretical position, provides an inconsistent and underdeveloped account of the mechanism for (phenomenal) consciousness. Despite motivating the Neural Stance with the claim that the location of RP is irrelevant to ascriptions of phenomenal consciousness, Lamme argues that only local instances of RP generate phenomena consciousness, while global RP generates mere access to consciousness. It is also unclear if Lamme can sustain the distinction between local RP with early pre-attentional phenomenal consciousness and global RP with late, cognitive access to consciousness. Global RP does more than provide mere access to consciousness; the occurrence, strength and type of content expressed in instances of local RP depend on earlier instances of global RP. Therefore the Neural Stance does not offer the kind of easy distinctions between types of neural processing that Lamme supposes.

The Neural Stance also commits Lamme to endorsing a view of conscious content that he, at least in his current discussions of the topic, does not explicitly acknowledge. Local RP, if it is identified with phenomenal consciousness, provides a range of multi-level, competing and inconsistent representations of stimulus information. In contrast, Lamme continues to endorse the kind of phenomenological descriptions found in traditional accounts of conscious content; those of single, detailed high level representations of stimuli. The fact that local RP provides far more phenomenal content than is currently acknowledged by Lamme may not be a problem if he agrees to bite the bullet. However, he has not done so to date, and it is not clear, given the kind of descriptions of phenomenal content he has used so far, if he would be willing to do so. The Neural Stance may be a productive one in terms of identifying natural distinctions in neural processing, but it must be properly applied, and when done so, may not identify distinctions that can be equated with those between phenomenal and access consciousness.

Both accounts also fail in terms of identifying satisfactory functions of consciousness. Attempting to functionalise consciousness in terms of reportability is no more precise than attempting to functionalise consciousness in terms of enabling subjects to succeed, indexed in a number of different ways, at a variety of tasks. Thus stated, reportability and related notions of 'access' are incredibly vague and fail to pick out a single target phenomenon. They instead refer to a vast range of phenomena that are produced by a range of task-specific mechanistic components. More generally, the top-down approach in consciousness science seems unlikely to succeed in describing an appropriate, high-level, target phenomena that is distinct from those already described within cognitive neuroscience. Bottom-up attempts to identify distinctions in low-level phenomena with those of conscious/unconscious or phenomenal/access consciousness are also unlikely to succeed in providing convincing equations between low-level mechanisms and consciousness. This is because it is becoming increasingly clear that neurobiological systems have very different properties than those attributed to consciousness. (This claim is developed throughout the next two chapters).

In effect, the same criticism has been levelled at both Lamme and GNWT; there is no simple way of decomposing all cases of complex, task-specific processing to support the distinction between conscious processing/reportability and background conditions (GNWT) or between phenomenal and access consciousness (Lamme), that respects normal standards of scientific practice.

Perhaps more clearly than the correlates and markers of consciousness discussed in the previous chapter, proposed mechanisms of consciousness are easily characterised as (parts of) mechanisms of well-known cognitive and neural phenomena. In the cognitive sciences these phenomena are recognised as being different, and generated by quite different mechanisms. They are however unceremoniously lumped together in consciousness science and expected to form a coherent, and scientifically interesting, whole. It is an empirical question where the scientifically interesting distinctions can be found, and it has been argued that they cannot be found using the conceptual framework provided by consciousness science.

In summary, as stated many times in earlier chapters, the distinctions we often refer to in relation to consciousness are often not scientifically valid ones, failing to be supported either by the application of experimental methods, or by experimental results. When scientific methods are appropriately applied to the investigation of these sorts of distinctions, they pick out a different range of phenomena from that originally intended, often with different boundaries, with no obvious or methodologically acceptable relation to 'consciousness'. In particular, it has been argued that the mechanisms proposed in consciousness science either fail to pick out a single phenomenon and scientific kind, leading to a lot of unproductive debates about finding the *real* mechanism of consciousness, or they pick out scientific kinds already investigated in cognitive neuroscience, re-labelled in a confusing and unproductive way. Current consciousness science therefore offers no new scientific kinds to research, and as suggested from the discussion above, obscures the products of existing and current research.

6.7 Conclusions

The mechanistic approach has lent further force to the claim of the previous chapter that consciousness science is not in fact a science of 'consciousness'. Earlier arguments were based on an analysis of integrative techniques, according to which groups of measures and theories of consciousness should be used to mutually refine one another. It was argued that the range of experimental protocols used, and the convergence between measures so far, shows that many different and distinct phenomena were being measured. In this case, measures of consciousness could not be productively integrated. Crucially, it was claimed that these phenomena were not different types or aspects of the same common phenomenon (consciousness).

By applying a different set of evaluative criteria to proposed mechanisms of consciousness, further arguments have been put forward to support this claim. Proposed mechanisms of consciousness, on closer inspection, are mechanisms for a range of disparate phenomena that are best left described as disparate phenomena. There also seems little to support the approach of dividing up earlier and later stages of task-specific processes into mechanisms of unconscious/conscious processing, or phenomenal/access conscious processing, given criteria of mechanistic demarcation. These processes can instead be decomposed into mechanisms that fulfill more specific functions related to types of information processing, stimulus-specific processing and decision-making, and so on. According to the HPC account of scientific kinds, in which mechanisms constitute kinds, the lack of well-demarcated and suitably targeted mechanisms of consciousness means that there are no scientific kinds that 'consciousness' refers to. In this case, 'consciousness' is not a viable scientific concept.

Before elaborating on this eliminativist claim, a final strategy for a science of consciousness will be discussed. This is the strategy of content-matching, of attempting to localise those brain areas that contribute to the contents of consciousness. Content-matching is analysed in the following two chapters, where it will be ultimately argued that this strategy is also highly problematic as it fails to reflect the current state of cognitive science. In this case, identity claims between conscious content and neural activity cannot be sustained. Along with criticisms of the methodology of consciousness science in other chapters, this seriously compromises the concept of 'consciousness' as a proper subject for scientific investigation.

References

Afraz, S. R., Kiani, R., & Esteky, H. (2006). Microstimulation of inferotemporal cortex influences face categorization. *Nature, 442,* 692–695.

Antal, A., Nitsche, M. A., Kruse, W., Kinces, T. Z., Hoffman, K.-P., & Paulus, W. (2004). Direct current stimulation over V5 enhances visuomotor coordination by improving motion perception in humans. *Journal of Cognitive Neuroscience, 16,* 521–527.

Bechtel, W. (2008). *Mental mechanisms: Philosophical perspectives on cognitive neuroscience.* London: Routledge.

Boehler, C. N., Schoenfeld, M. A., Heinze, H.-J., & Hopf, J.-M. (2008). Rapid recurrent processing gates awareness in primary visual cortex. *Proceedings of the Scientific Academy of Sciences, 105,* 8742–8747.

Boyd, R. (1989). What realism implies and what it does not. *Dialectica, 43,* 5–29.

Boyd, R. (1991). Realism, anti-foundationalism and the enthusiasm for natural kinds. *Philosophical Studies, 61,* 127–148.

Boyd, R. (1997). Kinds as the 'workmanship of men': Realism, constructivism, and natural kinds. In J. Nida-Rumelin (Ed.), *Rationality, realism, revision: Proceedings of the 3rd international congress of the Society for Analytical Philosophy* (pp. 52–89). New York: Walter de Gruyter.

Boyd, R. (1999). Homeostasis, species, and higher taxa. In R. A. Wilson (Ed.), *Species new interdisciplinary essays* (pp. 141–185). Cambridge: MIT Press.

Britten, K. H., & Wezel R. J. van. (1998). Electrical microstimulation of cortical area MST biases heading perception in monkeys. *Nature Neuroscience, 1,* 59–63.

Chalmers, D. J. (2000). What is a neural correlate of consciousness? In T. Metzinger (Ed.), *Neural correlates of consciousness.* Cambridge: MIT Press.

Craver, C. F. (2007). *Explaining the brain: Mechanisms and the mosaic unity of neuroscience.* Oxford: Clarendon Press.

Craver, C. F. (2009). Mechanisms and natural kinds. *Philosophical Psychology, 22*(5), 575–594.

Crick, F., & Koch, C. (1990). Towards a neurobiological theory of consciousness. *Seminars in the Neurosciences, 2,* 263–275.

Dehaene, S., & Changeux, J. P. (2004). Neural mechanisms for access to consciousness. In M. Gazzaniga (Ed.), *The cognitive neurosciences* (3rd ed., pp. 1145–1157). New York: Norton.

Dehaene, S., & Naccache, L. (2001). Towards a cognitive neuroscience of consciousness: Basic evidence and a workspace theory. *Cognition, 79,* 1–37.

Dehaene, S., Changeux, J. P., Naccache, L., Sackur, J., & Sergent, C. (2006). Conscious, preconscious, and subliminal processing: A testable taxonomy. *Trends in Cognitive Sciences, 100,* 204–211.

Del Cul, A., Baillet, S., & Dehaene, S. (2007). Brain dynamics underlying the nonlinear threshold for access to consciousness. *PLoS Biology, 5*(10), e260.

Griffiths, P. E. (1997). *What emotions really are: The problem of psychological categories.* Chicago: University of Chicago Press.

He, S., Cohen, E. R., & Hu, X. (1998). Close correlation between activity in brain area MT/V5 and the perception of a visual motion aftereffect. *Current Biology, 8,* 1215–1218.

Hess, R. H., Baker, C. L., Jr., & Zihl, J. (1989). The 'motion-blind' patient: Low-level spatial and temporal filters. *The Journal of Neuroscience, 9,* 1628–1640.
Hulme, O. J., Friston, K. F., & Zeki, S. (2008). Neural correlates of stimulus reportability. *Journal of Cognitive Neuroscience, 21,* 1602–1610.
Koivisto, M., & Revonsuo, A. (2003). An ERP study of change detection, change blindness, and visual awareness. *Psychophysiology, 40,* 423–429.
Koivisto, M., Revonsuo, A., & Salminen, N. (2005). Independence of visual awareness from attention at early processing stages. *NeuroReport, 16,* 817–821.
Koivisto, M., Revonsuo, A., & Lehtonen, M. (2006). Independence of visual awareness from the scope of attention: An electrophysiological study. *Cerebral Cortex, 16,* 415–424.
Koivisto, M., Railo, H., Revonsuo, A., Vani, S., & Salminen-Vaparant, N. (2011). Recurrent processing in V1/V2 contributes to categorization of natural scenes. *The Journal of Neuroscience, 31,* 2488–2492.
Kornblith, H. (1993). *Inductive inference and its natural ground: An essay in naturalisatic epistemology*. Cambridge: MIT Press.
Kouider, S., Dehaene, S., Jobert, A., & Le Bihan, D. (2007). Cerebral bases of subliminal and supraliminal priming during reading. *Cerebral Cortex, 17,* 2019–2029.
Lamme, V. A. F. (2004). Separate neural definitions of visual consciousness and visual attention; a case for phenomenal awareness. *Neural Networks, 17,* 861–872.
Lamme, V. A. F. (2006). Towards a true neural stance on consciousness. *Trends in Cognitive Sciences, 10,* 494–501.
Laureys, S. (2005). The neural correlates of (un)awareness: Lessons from the vegetative state. *Trends in Cognitive Sciences, 9,* 556–559.
Laureys, S., Owen, A. M., & Schiff, N. D. (2004). Brain function in coma, vegetative state, and related disorders. *The Lancet Neurology, 3,* 537–546.
Lee, T. S. (2002). Top-down influence in early visual processing: A Bayesian perspective. *Physiology and Behaviour, 77,* 645–650.
Lee, H. W., Hong, S. B., Seo, D. W., Tae, W. S., & Hong, S. C. (2000). Mapping of functional organization in human visual cortex: Electrical cortical stimulation. *Neurology, 54,* 849–854.
Logothetis, N., & Schall, J. (1989). Neuronal correlates of subjective visual perception. *Science, 245,* 761–763.
Machamer, P. K., Darden, L., & Craver, C. F. (2000). Thinking about mechanisms. *Philosophy of Science, 67,* 1–25.
Melloni, L., Molina, C., Pena, M., Torres, D., Singer, W., & Rodriguez, E. (2007). Synchronization of neural activity across cortical areas: Correlates with conscious perception. *The Journal of Neuroscience, 27,* 2858–2865.
Moutoussis, K., Keliris, G., Kourtzi, N., & Logothetis, N. (2005). A binocular rivalry study of motion perception in the human brain. *Vision Research, 45,* 2231–2243.
Mumford, D. (1992). On the computational architecture of the neocortex. II. The role of cortico-cortical loops. *Biology and Cybernetics, 66,* 241–251.
Murphey, D. K., Maunsel, J. H. R., Beauchamp, M. S., & Yoshor, D. (2009). Perceiving electrical stimulation of identified human visual areas. *Proceedings of the National Academy of Sciences, 106,* 5389–5393.
Nawrot, M., & Rizzo, M. (1988). Chronic motion perception deficits from midline cerebellar lesions in human. *Vision Research, 38,* 2219–2224.
Newsome, W. T., & Paré, E. B. (1988). A selective impairment of motion perception following lesions of the middle temporal visual area (MT). *The Journal of Neuroscience, 8,* 2201–2211.
Nichols, M. J., & Newsome, W. T. (2002). Middle temporal visual area microstimulation influences veridical judgements of motion direction. *The Journal of Neuroscience, 22,* 9530–9540.
Palva, S., Linkenkaer-Hansen, K., Näätänen, R., & Plave, J. M. (2005). Early neural correlates of conscious somatosensory perception. *The Journal of Neuroscience, 25,* 5248–5258.
Pasternak, T., & Merigan, W. H. (1994). Motion perception following lesions of the superior temporal sulcus in the monkey. *Cerebral Cortex, 4,* 247–259.

Sadaghiani, S., Scheeringa, R., Lehongre, K., Morillon, B., Giraud, A.-L., & Kleinschmidt, A. (2010). Intrinsic connectivity networks, alpha oscillations, and tonic alertness: A simultaneous electroencephalography/functional magnetic resonance imaging study. *The Journal of Neuroscience, 30,* 10243–10250.
Salzman, C. D., Murasugi, C. M., Britten, K. H., & Newsome, W. T. (1992). Microstimulation in visual area MT: Effects on direction discrimination performance. *The Journal of Neuroscience, 12,* 2331–2355.
Scholte, H. S., Witteveen, S. C., Spekreijse, H., & Lamme, V. A. F. (2006). The influence of inattention on the neural correlates of scene segmentation. *Brain Research, 1076,* 106–115.
Scholte, H. S., Jolij, J., Fahrenfort, J. J., & Lamme, V. A. F. (2008). Feedforward and recurrent processing in scene segmentation: Electroencephalography and functional magnetic resonance imaging. *Journal of Cognitive Neuroscience, 20,* 2097–2109.
Sergent, C., & Dehaene, S. (2004a). Is consciousness a gradual phenomenon? Evidence for an all-or-none bifurcation during the attentional blink. *Psychological Science, 15,* 720–728.
Sergent, C., & Dehaene, S. (2004b). Neural processes underlying conscious perception: Experimental findings and a global neuronal workspace framework. *Journal of Physiology, 98,* 374–384.
Sergent, C., Baillet, S., & Dehaene, S. (2005). Timing of the brain events underlying access to consciousness during the attentional blink. *Nature Neuroscience, 8,* 1391–1400.
Shea, N., & Bayne, T. (2010). The vegetative state and the science of consciousness. *British Journal for the Philosophy of Science, 61,* 459–484.
Supèr, H., Togt, C. Van Der, Spekreijse, H., & Lamme, V. A. F. (2003). Internal state of monkey primary visual cortex (V1) predicts figure-ground perception. *Journal of Neuroscience, 23,* 3407–3414.
Wilson, R. A. (1999). Realism, essence, and kind: Resuscitating species essentialism? In R. A. Wilson (Ed.), *Species: New interdisciplinary essays* (pp. 187–207). Cambridge: MIT Press.
Zeki, S. M. (1990). The motion pathways of the visual cortex. In C. Blakemore (Ed.), *Vision: Coding and efficiency* (pp. 321–345). Cambridge: Cambridge University Press.

Chapter 7
Content-Matching: The Case of Sensory Memory and Phenomenal Consciousness

7.1 Introduction

Despite the problems in using dissociation, integration and mechanistic methods to investigate consciousness, there is another alternative strategy to use in consciousness research. This is to try to match the contents of consciousness with the contents of particular stages of information processing in the brain. One way to do this is to look for neural correlates of consciousness (NCCs) that map neural representational contents to the contents of consciousness (partly discussed in the previous chapter). A different way of content-matching is to try to find similarities between the apparent properties of (phenomenal) consciousness and the properties of particular stages of perceptual processing. If a stage of processing provides the kind of content that matches a phenomenological description of perceptual content, then this stage of processing could be identified with consciousness of this content.

To pursue this strategy, it is first necessary to give a description of the contents of consciousness and their properties. The attempt at content-matching discussed here is based on the 'rich' view of the contents of consciousness. This view suggests that pre- or unattended contents, while not accessed or reported, are still present in the contents of consciousness. It is argued that unreported visual information is experienced on a different 'level' to reported visual information. Information about attended and reported objects/areas is 'cognitively accessed' (Block 2001, 2005, 2007), and is experienced on a conceptual (Tye 2006) or 'fact' (Dretske 2004, 2007) level of awareness. In contrast, information about unattended and unreported objects/areas is not cognitively accessed, but is experienced on a different 'phenomenal' (Block), non-conceptual (Tye), or 'object' (Dretske) level of awareness.

The rich view therefore suggests a particular view of the contents of consciousness. Proponents of this view have also suggested how to locate the rich contents of consciousness within sensory processing, typically by referring to the results of

This chapter is a version of 'Rich experience and sensory memory', E. Irvine, Philosophical Psychology, Vol. 24 Issue 2, pp. 159–176, Copyright 2011. Reprinted with permission from Taylor and Francis Ltd.

E. Irvine, *Consciousness as a Scientific Concept*, Studies in Brain and Mind 5,
DOI 10.1007/978-94-007-5173-6_7, © Springer Science+Business Media Dordrecht 2013

partial report paradigms (e.g. Sperling 1960). These paradigms show that more information than subjects can concurrently report is stored in a form of iconic or sensory memory. As this is consistent with the idea of phenomenology 'overflowing' cognitive access, many authors have recently claimed that the contents of sensory memory *are* the contents of phenomenal experience (see e.g. Block 2007; Fodor 2007, 2008; Jacob and De Vignemont 2010; Tye 2006). This chapter examines the equation made between the 'rich' contents of phenomenal consciousness and the contents of sensory memory. Taken as a case study of the content-matching strategy, this examination focuses on two main questions.

The first question is whether the contents of sensory memory are of an appropriate form to be equated with the contents of phenomenal consciousness. It will be argued that sensory memory does not provide the sort of static, unitary and detailed and pre-categorical representation that is necessary for the way the rich view describes the contents of phenomenal consciousness. The second question concerns the way in which the contents of phenomenal consciousness are identified by the rich view. Along with an interpretation of Sperling's paradigm, the rich view also depends on a particular interpretation of subjects' reports of what they experience. The second question therefore focuses on how much detailed visual information is necessary to generate reports of rich visual experiences.

It will be argued that the answers to both of these questions seriously undermine the currently popular claim that the contents of sensory memory are the contents of phenomenal consciousness. They further provide evidence that the current neuropsychological understanding of the visual system is simply not consistent with a taxonomy of phenomenal and accessed conscious states, and instead supports a radically different taxonomy of cognitive abilities. An alternative model that explains both task performance and report generation is given in the following chapter, through which the strategy of content-matching is shown to be based on the misconception that 'the contents of consciousness' can refer to a unified set of uni-level representations.

7.2 Partial Report Superiority

In order to claim that detailed but unreported representations of scenes form part of the contents of consciousness, supporters of the rich view often cite the results of partial report paradigms (Landman et al. 2003; Sligte et al. 2008; Sperling 1960). Aside from appeals to intuition, this paradigm provides the primary evidence in favour of the rich view and its corresponding taxonomy of phenomenal and accessed visual consciousness. The paradigm suggests that a short term, large capacity sensory memory stores detailed visual information. Only some of this information can be accessed via attentional selection, and most of this information is quickly 'forgotten'. However, it is argued that this unaccessed content could provide subjects with a rich set of phenomenally conscious content. The original paradigm and modern variations are described below.

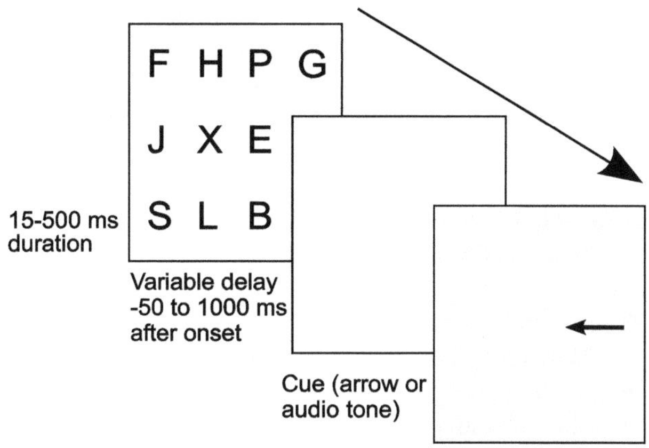

Fig. 7.1 Partial report paradigm (full report condition does not include the cue)

Sperling (1960) claimed to establish the existence of a pre-categorical 'iconic' memory store through the phenomenon of partial report superiority. In his classic paradigm, subjects are shown a display of letters for a short time (15–500 ms), followed by a variable delay, and asked to report as many letters as they can. Subjects in this 'full report' condition are capable of reporting around 4.5 letters out of 12 letters, which is reasonably invariant to the duration of the display. In the partial report condition, a display of letters are again shown for a short time, and now a cue for a particular row is given some time before or after the stimulus (e.g. from 50 ms before onset up to 1,000 ms after stimulus offset). The cue can either be visual, such as an arrow, or an audio cue, with high, medium and low tones cuing top middle and bottom rows respectively. After the cue there is a further variable delay before the subject responds. In this condition, subjects report as many letters as they can from the cued row, and on average report 3.03 letters from each four item line (Fig. 7.1).

Sperling suggested that since subjects reported this number of letters *for any cued row*, information must be available for about 9.01 items from the 12 letters, far higher than is suggested in the full report condition. Subjects also report seeing all the letters in the display, despite not being able to report all of their identities. Sperling suggested that a form of short term 'iconic' memory would explain these results. Iconic memory would store large amounts of low level (pre-categorical) information for brief periods of time, but only a small amount of it would be fully processed and reported at any one time. This explains how information from any cued row could be accessed, but the short duration of this form of memory ensures that information from non-cued rows would decay before subjects could report it.

More recent versions of this paradigm suggest that visual short term memory has a much larger capacity and longer duration than Sperling found. Landman et al. (2003) and Sligte et al. (2008) used experimental paradigms that combine aspects of the Sperling and change blindness paradigms. Landman et al. showed subjects eight differently oriented (and differently oriented *and* sized) rectangles for 500 ms, followed by a delay, followed by a display in which one of the rectangles had changed. Again, in the full report condition subjects showed poor performance at identifying the changed rectangle, consistent with other change blindness paradigms. However, when cues that mark particular rectangles are shown between the initial and changed display, performance increases hugely. Subjects' performance in the cued condition suggests that with practice they can maintain information about six to seven out of eight rectangles over 1.5 s after stimulus offset. Sligte et al. (2008) took this paradigm further and used 32 oriented rectangles in a similar change detection task. Although the strength of after images and possible chunking effects (see Brockmole and Wang 2003) may play a significant role in determining their results, they also found evidence for a larger capacity and longer- lived form of sensory memory than found in the original paradigm (see also Phillips 2011 for alternative explanation).

The results of these partial report tasks are consistent with, and seem to provide empirical support for, the claims of the rich view. There is a kind of memory that contains large amounts of detailed information that is not all concurrently available, and is available for only short periods of time. Subjects' reports indicate that this information is experienced. The contents of this form of memory could therefore provide the contents for a richly populated level of consciousness. Supporters of the rich view have therefore claimed that the contents of sensory memory *are* the contents of phenomenal consciousness.

There are two ways in which this claim can be assessed. The first is whether the character of sensory memory is consistent with the existence of a rich and detailed level of phenomenal consciousness. It will be argued that the use of partial report superiority to support the rich view depends on a misunderstanding of what partial report tasks actually measure, and an outdated understanding of sensory memory. The second problem follows on from a clarification of what partial report paradigms are designed to investigate; the persistence of information in memory and not the contents of consciousness. Given this, it is only subjects' reports of visual richness that link empirical data from these paradigms with claims about the contents of experience. It will be argued that reports of visual richness can be generated even when visual details are not present in sensory memory. In this case, subjects' reports cannot be used as evidence to show that detailed visual information in sensory memory is identical with the contents of phenomenal consciousness.

7.3 The Structure of Sensory Memory: Visible and Informational Persistence

The original model of sensory memory used to explain the phenomenon of partial report superiority is as a form of 'iconic' memory. Iconic memory was characterised as a short-lived, pre-categorical memory from which limited amounts of information could be accessed at any one time. It is this model of iconic memory that supporters of the rich view use to argue for a full phenomenal experience of a display. In fact this characterisation of iconic memory became discredited by the 1980s as research showed that visual short term memory was not unitary, and memory research instead re-focused on the properties of the different forms of memory that together are now referred to as sensory memory (see Loftus and Irwin 1998; Luck and Hollingworth 2008). Instead of being a unified store of low-level pre-categorical information, sensory memory contains two stores of different kinds of information, incorporating several levels of visual processing, some of which is categorical in nature.

Current models of sensory memory are built to reflect two different phenomena of short term visual processing; visible and informational persistence. Although they are often confusingly identified as bringing about the same perceptual effects they are quite different. Visible persistence refers to activity in early visual areas that continues after a stimulus is presented, that typically enables the integration of visual information over short time scales. Visible persistence is typically investigated using temporal integration paradigms, in which spatial arrays are presented sequentially and superimposed (see e.g. Di Lollo 1980). Under certain presentation conditions, these arrays are seen as a single, integrated array.

After-images provide another example of visible persistence, and are the result of activity that persists for much longer than normal, for example as a result of adaptation followed by a high contrast stimulus (lightning against a dark sky). Visible persistence is simply the result of ongoing stimulus-specific neural activity in early visual areas, that affects how stimuli are processed, and in some cases how long they are visually present. Although part of sensory memory, visible persistence is not really a type of memory at all.

In contrast, informational persistence refers to later stages of visual processing, and is the aspect of sensory memory that gives rise to partial report superiority. Informational persistence refers to a short lived but high (higher than working memory) capacity, short term type of memory under which information is stored for several hundred milliseconds. As shown in partial report paradigms, some of this information can be accessed and reported but most is immediately 'forgotten'. Information persists in two different ways; in a visible analogue representation, and in a post-categorical store. The visible analogue representation preserves shape and location information and is stored for 150–300 ms after stimulus offset. The post-categorical store preserves abstract information such as identity for up to 500 ms. In partial report tasks, identity information from the post-categorical store is matched with location information from the visible analogue representation in order to report letter identities from cued rows.

Partial report paradigms are used by supporters of the rich view to identify the contents of consciousness with the contents of persisting visual experiences. However, partial report paradigms are only capable of assessing the properties of persisting information: "the partial-report technique does not measure directly the *visible* aspect of visual sensory memory, but rather that *information* persists after stimulus onset" (Luck and Hollingworth 2008, p. 16, original italics). Partial report superiority is a memory phenomenon concerning the quantity of information that can be stored and reported from brief visual displays. As a phenomenon concerned purely with short-term information recall, it has no clear relevance to discussions about the contents of consciousness, either of what the contents are or how long they are conscious for. Whether the stored information that is probed in partial report paradigms is consciously experienced is not something that can be inferred from these paradigms themselves (as paradigms concerning memory only), or from temporal integration paradigms (as they assess properties of much earlier visual processing). Several problems in using partial report paradigms to identify the rich contents of consciousness are discussed below.

7.3.1 One Store, Two Stores

That visible and informational persistence are different phenomena, occurring at different stages of visual processing, is largely ignored by many supporters of the rich view. From this they erroneously assume that visible and informational persistence are different aspects on the same phenomenon, and thus that partial report superiority *is* relevant to identifying the contents of consciousness. For example, Block clearly confuses the two types of persistence. In reference to Landman et al. (2003) he states: "Subjects are apparently able to hold the visual experience for up to 1.5 s—at least 'partial report superiority' (as it is called) lasts this long" (Block 2007, p. 488). Partial report tasks do *not* tell us anything about how long experiences last, only how long information persists in memory. Empirical data from partial report tasks, by itself, does not contain any information about the phenomenal states subjects experience when they view displays; partial report superiority is consistent with subjects having no experience at all.

Tye (2006) also makes unwarranted conclusions about how the contents of informational persistence contribute to the contents of consciousness. While he acknowledges that partial report paradigms examine the availability of *information*, he subsequently concludes that this information "operates at the phenomenal level" (p. 511). Tye refers to temporal integration tasks and visual after images to argue that visual information is sometimes experienced for longer than information is actually visually present. From this, he concludes that the unaccessed contents of sensory memory identified in partial report paradigms, which can also be present after stimulus offset, are also experienced. However, the properties of visible persistence, exhibited in temporal integration tasks, cannot be applied to the unaccessed contents of sensory memory identified in partial report task. Visible persistence and informational persistence are two very different phenomena that arise at different stages of visual processing.

This mistake of assuming that visible and informational persistence refer to different aspects of the same (conscious) phenomenon seems to be a common one. Instead, sensory memory is an umbrella term, referring to different kinds of early visual processing. Paradigms used to investigate visible persistence are simply irrelevant when assessing whether persisting and unreported information in later visual areas forms part of the contents of consciousness. Partial report paradigms are concerned with the informational contents of visual short-term memory, including separate stores of location and identity information. This fragmentation of sensory memory also makes it difficult to refer to 'the' contents of sensory memory. It is consistent with Block and Tye's usage for these contents to stretch from very early visual activity (visible persistence) to areas involved in object identification (informational persistence).

7.3.2 *Is Sensory Memory Iconic?*

Another crucial point in assessing the identity claim between the contents of sensory memory and the contents of phenomenal consciousness is to establish how well the properties of the contents of sensory memory match the properties of the contents of consciousness. It will be argued that sensory memory does not generate or preserve an 'iconic' or pre-categorical representation of the kind that can be matched to subjects' reports about the paradigm (for example, that subjects report 'reading off' letter identities from a representation of an array, Shanahan and Baars 2007). Sensory memory does not consist of a store of unprocessed visual information that is experienced by subjects prior to being accessed and reported. Instead, sensory memory refers to a set of dynamically changing and deeply processed fragmented set of information in the visual system.

First, the way that different kinds of persisting information develop and decay in sensory memory illustrates difficulty in describing it as 'iconic' or as a single, static visual representation. For example, item specific processing starts around 100 ms with letter-related processing occurring around 150 ms, and high-level case independent representations present after 220 ms (Tarkiainen et al. 2002; Thorpe et al. 1996). This information is stored in the post-categorical store, and decays between 300–500 ms after stimulus offset. This contrasts with the different decay rate of the visible analog store, which ends 150–300 ms after stimulus offset.

For cues given between 300 and 500 ms after stimulus offset, the visible analogue representation (location information) has decayed, but the post-categorical store (letter identity) is still active. When using these cues subjects systematically make 'location errors' in their responses. They can correctly identify some letters from the display, but subjects are no longer able to match letter identities with particular rows as location information is no longer available. Clearly, being able to access letter identities without letter locations is not consistent with the idea of 'reading off' letters from a visual experience or even a pre-categorical visual store, as suggested by the rich view.

From spatial processing and letter identification, and through different decay trajectories, the contents of sensory memory radically change over short time scales. However, subjects do not report a change in experience that tracks the difference in content that these different decay rates determine. It is not the case that location information somehow disappears from visual consciousness before letter identity information does. While the contents of sensory memory are the sequential products of different levels of visual processing, the contents of visual experience are not. Given that sensory memory does not provide the kind of stable representations that map onto what subjects claim to experience, there is a significant problem in claiming that contents of sensory memory are the contents of phenomenal experience.

7.3.3 Visual Detail in Natural Scenes

Even if a large amount of item-specific visual information is available in sensory memory, partial report paradigms only suggest this that is true for a very particular kind of stimulus array. The letter and rectangle displays used in partial report tasks are designed such that identification of items is easy. However, the conclusions that supporters of the rich view draw from these kinds of displays cannot be easily generalised to perception of natural scenes. Although some authors are aware of this (e.g. Tye 2006), it is not a generally acknowledged problem.

The letters and rectangles used in partial report tasks are sparsely distributed, and presented within an easily foveated, fixated area for a short period of time. Natural scene perception operates under very different conditions, where several perceptual effects can be observed that are eradicated when using sparse displays. When viewing displays that have many stimuli close together, perceptual crowding and lateral masking prevent easy identification or discrimination of stimuli (e.g. see Pelli et al. 2004; Wertheim et al. 2006). These two phenomena play in a role in the way in which information from visual scenes is processed and experienced, yet are left out of accounts based on the perception of sparse arrays only.

In fact, there is some evidence that partial report superiority is not found in natural scene perception. In contrast with the results of partial report tasks used by Landman et al. (2003) and Sligte et al. (2008); Liu and Jiang (2005) found that using a comparable recognition task for items in a natural scene, performance was very low (0.7 items recognized compared to up to 32 items in Sligte et al. 2008). The authors comment that "although 250 ms may be sufficient for people to perceive semantic gist of a scene . . . it is not long enough for most visual details to enter VWM [i.e., to be reported post-cue]" (Liu and Jiang 2005, p. 656).

Therefore there are reasons to suppose that the phenomenon of partial report superiority for sparse and simple arrays cannot be used to argue for the existence of a richly detailed level of experience for natural scenes. The constraints on visual processing of larger and more complex scenes suggest that unreported and unattended visual information is not always available in the same detailed format as for simpler arrays.

7.3.4 Summary

It has been argued that partial report paradigms do not support an equation between the contents of rich conscious content and the content of sensory memory. Partial report paradigms assess informational persistence only, a memory phenomenon, and cannot be used to identify the contents of consciousness. Further, it was argued that sensory memory does not provide the kind of unitary static pictorial representations that supporters of the rich view assume. Different kinds of information (spatial/categorical) are stored in different ways and decay at different rates. Finally, it was also argued that generalising from results gained in partial report tasks to perception in general is also problematic. The following section focuses on the different question of the relation between reports of visual richness and the contents of visual processing.

7.4 Interpretations and Reports

It was argued above that partial report paradigms provide evidence only about the persistence of information in visual short-term memory. However, these experiments seem relevant to discussions of the contents of consciousness because subjects also give phenomenological reports about their experiences. The fact that subjects report seeing *all* the letters or *all* the rectangles provides the essential link in arguing that the unreported contents of sensory memory are experienced on a rich phenomenal level. This essential link requires investigation to see if the presence of rich representations in the visual system is necessary to generate reports of visual richness. If they are not, this further undermines the equation between the contents of phenomenal consciousness with the contents of sensory memory, and calls into question how reports of richness are often interpreted.

Various authors have offered conceptual arguments to show that experiences and reports of richness do not depend on the presence of rich information in the brain. Block himself calls it the 'photographic fallacy' (2007 p. 533), Van Gulick refers to it as the 'movie screen of the mind' model of consciousness (2007 p. 529), and Dennett has also extensively argued against the idea of the 'Cartesian theatre' (1991 esp. pp. 354–355). However, there is also a growing amount of empirical work to suggest that processing and storing detailed information is not necessary for the generation of reports of experiencing, knowing, or remembering detailed information. In recent years the large role played by the processing and storing of gist and high level categorical information has been recognised, both in vision and memory research.

Beginning with research into memory, the Deese-Roediger-McDermott (DRM) paradigm (see Roediger and McDermott 1995) shows how strong encoding of the theme of a word list leads to high false recognition rates for words consistent with the list but not actually on it. In the visual domain, scene-level information (gist) can be processed from stimulus durations as short as 20 ms, and can be completed in less than 100 ms from stimulus onset (see e.g. Oliva 2005; Van Rullen and Thorpe 2001). Research into scene gist in visual perception also shows how the specific items that subjects report from a scene are strongly influenced by the meaning of the scene. For example, Brewer and Trevans (1981) found that subjects tended to report seeing books when they had spent some time in an office, even though none were present.

This response bias towards (sometimes falsely) recognizing objects consistent with the gist of a scene is a widely found and reliable phenomenon (see e.g. Henderson Hollingworth 1999; Hollingworth and Henderson 2002; Pezdek et al. 1989), and is utilized in Castelhano and Henderson's (2008) contextual bias paradigm. Subjects are shown a photograph of a scene for 20–250 ms followed by a 50 ms mask, and then asked whether a target object is present in the photograph. For a scene of a park, a target object might be a bench (consistent) or a refrigerator (inconsistent). Subjects tend to respond that consistent items are present and that inconsistent items are not, despite the fact that *none* of the target objects are actually present in the scenes. Castelhano and Henderson note:

> ...participants in this experiment were not told until debriefing that the objects were not actually present in any of the scenes. When asked during the debriefing period whether they had suspected this during the experiment, participants typically reported that they had not noticed. (2008, p. 665)

This extreme example of the way in which subjects rely on scene gist shows how reports about specific objects and visual detail often fail to reflect any processing of detailed visual information. Of particular relevance to the reports of visual richness generated in partial report paradigms, de Gardelle et al. (2009) tested the role of scene-level information and expectation in response generation for simple letter arrays. When shown a display of letters with pseudo-letters in non-cued rows, subjects still report seeing 'all the letters,' and are surprised when told that displays contain these pseudo-letters. Scene-level category information (that the items are letter-like) combined with expectations about the contents of the display drive responses about the presence of specific objects, sometimes misleading subjects who give an incorrect description of the contents of a scene (see also Kouider and Dupoux 2004).

Partial report paradigms are used to show the existence of a memory store of richly detailed information, but these reports of visual richness are clearly not based on the detailed item-specific contents of this store. Instead, scene-level information plays the primary role in determining the kinds of reports that are later used as evidence that detailed persisting information is experienced on a phenomenal level. The evidence above shows that reports of visual richness cannot be used in this way, and fail to provide support for the existence of a richly detailed level of unaccessed, phenomenal experience.

7.5 Conclusion

It has been argued that neither the phenomenon of partial report superiority, nor phenomenological reports of visual richness, support the claim that there is a rich and detailed set of unreported conscious content. Partial report superiority provides evidence only about the persistence of information in sensory memory and in itself says nothing about the contents of consciousness. The generation of reports of seeing 'all the letters' in a display, and reports of visual richness in general, do not rely on the processing of rich detailed visual information. Although larger amounts of information are present in sensory memory than can be concurrently reported, subjects' reports of visual richness are not sufficient to show that these contents contribute to the contents of consciousness. This attempt at content-matching clearly fails.

A description of a research heuristic commonly used when proposing identity claims in science, McCauley's and Bechtel's Heuristic Identity Theory (2001), suggests that identity claims are used to make testable predictions, and to revise the concepts used at both levels of the identity claim in light of these tests. The problems with the identity claim examined here suggest that serious reconceptualisation is required if any identity can be supported between the contents of consciousness (as described above) and the contents of information processing. This will include the rejection of the assumption that the unreported contents of perception are not processed as deeply as content that is reported, and that the contents of consciousness are supplied by a single representation.

The following chapter takes this investigation further; of how descriptions of the contents of consciousness differ from properties of visual processing. This will be done by examining cases where subjects' reports and behaviours towards the same stimuli appear inconsistent. Again, the properties attributed to conscious content will be shown to be quite different to those of sensory and cognitive content. In particular, it will be shown that there are serious problems in trying to describe a set of content that is demarcated in an empirically plausible way, and that satisfies the phenomenal properties typically attributed to conscious content. It will be argued that far from providing a viable research program, advancing empirically informed accounts of the contents of consciousness is deeply problematic.

References

Block, N. (2001). Paradox and cross purposes in recent work on consciousness. *Cognition, 79*, 197–219.

Block, N. (2005). Two neural correlates of consciousness. *Trends in Cognitive Science, 9*, 46–52.

Block, N. (2007). Consciousness, accessibility, and the mesh between psychology and neuroscience. *Behavioral and Brain Sciences, 30*, 481–548.

Brewer, W. F., & Trevans, J. C. (1981). The role of schemata in memory for places. *Cognitive Psychology, 13*, 207–230.

Brockmole, J. R., & Wang, R. F. (2003). Integrating visual images and visual percepts across space and time. *Visual Cognition, 10*, 853–873.

Castelhano, M. S., & Henderson, J. M. (2008). The influence of color on the activation of scene gist. *Journal of Experimental Psychology: Human Perception and Performance, 34*, 660–675.

Gardelle, V. de, Sackur, J., & Kouider, S. (2009). Perceptual illusions in brief visual presentations. *Consciousness and Cognition, 18*, 569–577.

Dennett, D. C. (1991). *Consciousness explained.* London: Penguin.

Di Lollo, V. (1980). Temporal integration in visual memory. *Journal of Experimental Psychology: General, 109*, 75–97.

Dretske, F. (2004). Change blindness. *Philosophical Studies, 120*, 1–18.

Dretske, F. (2007). What change blindness teaches about consciousness. *Philosophical Perspectives, 21*, 215–230.

Fodor, J. A. (2007). The revenge of the given. In B. P. McLaughlin & J. Cohen (Eds.), *Contemporary debates in philosophy of mind* (pp. 105–116). Oxford: Blackwell.

Fodor, J. A. (2008). Preconceptual representation. *LOT2: The language of thought revisited* (pp. 169–196). New York: Oxford University Press.

Henderson, J. M., & Hollingworth, A. (1999). The role of fixation position in detecting scene chances across saccades. *Psychological Science, 10*, 438–443.

Hollingworth, A., & Henderson, J. M. (2002). Accurate visual memory for previously attended objects in natural scenes. *Journal of Experimental Psychology: Human Perception and Performance, 28*, 113–136.

Jacob, P., & De Vignemont, F. (2010). Spatial coordinates and phenomenology in the two-visual systems model. In N. Gangopadhyay (Ed.), *Perception, action, and consciousness: Sensorimotor dynamics and two-visual systems* (pp. 125–144). Oxford: Oxford University Press.

Kouider, S., & Dupoux, E. (2004). Partial awareness creates the "illusion" of subliminal semantic priming. *Psychological Science, 15*, 75–81.

Landman, R., Spekreijse, H., & Lamme, V. A. F. (2003). Large capacity storage of integrated objects before change blindness. *Vision Research, 43*, 149–164.

Liu, K., & Jiang, Y. (2005). Visual working memory for briefly presented scenes. *Journal of Vision, 5*, 650–658.

Loftus, G. R., & Irwin, D. E. (1998). On the relations among different measures of visible and informational persistence. *Cognitive Psychology, 35*, 135–199.

Luck, S. J., & Hollingworth, A. (2008). *Visual memory*. New York: Oxford University Press.

McCauley, R. N., & Bechtel, W. (2001). Explanatory pluralism and heuristic identity theory. *Theory and Psychology, 11*, 736–760.

Oliva, A. (2005). The gist of a scene. In L. Itti, G. Rees, & J. K. Tsotsos (Eds.), *The neurobiology of attention* (pp. 251–256). London: Academic Press.

Pelli, D. G., Palomares, M., & Majaj, N. J. (2004). Crowding is unlike ordinary masking: Distinguishing feature integration from detection. *Journal of Vision, 4*, 1136–1169.

Pezdek, K., Whetstone, T., Reynolds, K., Askari, N., & Dougherty, T. (1989). Memory for realworld scenes: The role of consistency with scheme expectation. *Journal of Experimental Psychology: Learning, Memory, and Cognition, 15*, 587–595.

Phillips, I. B. (2011). Perception and iconic memory. *Mind & Language, 26*, 381–411.

Roediger, H. L., & McDermott, K. B. (1995). Creating false memories: Remembering words not presented in lists. *Journal of Experimental Psychology: Learning, Memory and Cognition, 21*, 803–814.

Shanahan, M., & Baars, B. (2007). Global workspace theory emerges unscathed. *Behavioral and Brain Sciences, 30*, 524–525.

Sligte, I. G., Scholte, H. S., & Lamme, V. A. F. (2008). Are there multiple visual short-term memory stores? *PLoS 1, 3*, 1–9. doi:10.1371/journal.pone.0001699.

Sperling, G. (1960). The information available in brief visual presentations. *Psychological Monographs: General and Applied, 74*, 1–29.

Tarkiainen, A., Cornelissen, P. L., & Salmelin, R. (2002). Dynamics of visual feature analysis and object level processing in face versus letter-string perception. *Brain, 125*, 1125–1136.

Thorpe, S., Fize, D., & Marlot, C. (1996). Speed of processing in the human visual system. *Nature, 381*, 520–522.

Tye, M. (2006). Nonconceptual content, richness, and fineness of grain. In T. G. Szabo & J. Hawthorne (Eds.), *Perceptual experience* (pp. 504–530). Oxford: Oxford University Press.

Van Gulick, R. (2007). What if phenomenal consciousness admits of degrees? *Behavioral and Brain Sciences, 30*, 528–529.

Van Rullen, R., & Thorpe, S. (2001). The time course of visual processing: From early perception to decision making. *Journal of Cognitive Neuroscience, 13*, 454–461.

Wertheim, A. H., Hooge, I. T. C., Krikke, K., & Johnson, A. (2006). How important is lateral masking in visual search? *Experimental Brain Research, 170*, 387–402.

Chapter 8
Content-Matching: The Contents of What?

8.1 Introduction

The previous chapter examined the identity claim that has been made between the contents of phenomenal consciousness and the contents of sensory memory. The specific problems with this attempt at content-matching raise more general questions about the strategy of content-matching in consciousness science, explored in this chapter. While the rich view of conscious content was the subject of the previous chapter, three other accounts, including sparse, sensori-motor and hybrid accounts of the contents of consciousness are discussed here. The ways that these four accounts treat the relationship between reports, behaviours and conscious content are outlined below.

Many of these accounts attempt to explain the conflicting reports and behavioural evidence found in change and inattentional blindness and partial report superiority, but they do so in a way that is often inconsistent with more empirically informed models of the generation of reports and behaviours. Referring to a general model of the relationships between perception, reports and behaviours, in terms of multi-stream generative information processing, makes it possible to identify problems with the ways that these accounts isolate the set of conscious content. This material is then used to explore the ways in which all of the current accounts of conscious content, and the common assumptions underlying them, are incompatible with commonly accepted features of visual processing.

Throughout this chapter the appraisals of contemporary accounts of conscious content will not attempt to be comprehensive, nor will they attempt to be particularly novel. They are instead used to point to a number of examples in which a seemingly plausible identity claim between the contents of consciousness and the contents of perceptual processing break down, given a closer look at the concepts used at both levels of description. This means that identity claims made between conscious content and the content of processing streams all fail, either from a mismatch with empirical models (typically of visual perception), or from a mismatch with widely acknowledged features of conscious content. However, what is of real philosophical interest are the methodological options available in consciousness science once these mismatches have been identified.

E. Irvine, *Consciousness as a Scientific Concept*, Studies in Brain and Mind 5,
DOI 10.1007/978-94-007-5173-6_8,

McCauley and Bechtel's Heuristic Identity Theory (2001) will be used to suggest how the content-matching strategy fails to function in current consciousness science, and why that failure is an important methodological one. McCauley and Bechtel argue that identity claims, such as those found in the content-matching strategy, are initially put forward as hypotheses to be refined and reformulated given research at both levels of the identity. "Hypothesising cross-scientific identities is a pivotal engine of scientific development. Hypothetical identities in interlevel contents serve as valuable heuristics of discovery for inquiry at both of the explanatory levels involved" (McCauley and Bechtel 2001, p. 751). If an identity (or matching) claim fails, then this points to ways in which concepts at both levels need to be revised.

Using this framework it will be argued that the differences between the properties of sensory and cognitive content, and the assumed properties of contents of consciousness, force a serious reappraisal of the idea of 'the contents of consciousness' and the strategy of content-matching in general. The implications of the failure of the content-matching strategy in consciousness science, combined with the failures of the other strategies and methods discussed in earlier chapters, are examined in full in the next chapter.

8.2 Conflicting Contents: Reports vs. Behaviour

There are two main ways that consciousness is operationalised; some form of verbal report, or some other measurable behavioural response (e.g. forced choice response). However, in some experimental paradigms, even reports can identify apparently conflicting sets of conscious content. For example, in the Sperling, change blindness and inattentional blindness paradigms, subjects report being aware of a whole scene yet are only able to identify a few salient items at a time, or fail to notice salient changes or events altogether. This raises the question of which type of reports or responses are better indicators of the real contents of consciousness. These problematic cases have given rise to four main accounts that attempt to solve the conflict between scene-level and item-specific reports and thereby identify the 'real' contents of consciousness. These accounts are outlined briefly below and analysed in the following sections.

8.2.1 Richness/Internalism

Explored in the previous chapter, and endorsed by philosophers and some neuroscientists (e.g. Block 2007; Lamme 2006), the rich view takes reports at face value, concluding that the contents of consciousness identified in reports are all actual, not merely potential, contents. However, as subjects cannot use attended information in the same way as unattended information, this suggests that a two-stage model of perception is necessary. Consciousness of rich content (phenomenal consciousness) is followed by sparse informational uptake for further processing and report (access to consciousness).

While this model provides an account to accommodate both reports and behaviours, and explains why they sometimes identify conflicting contents, attempts to identify the locus of the first 'rich' stage of non-conceptual or phenomenally conscious content are problematic. The problems with Lamme's identification of phenomenal consciousness with local recurrent processing were discussed in Chap. 6. Attempts to identify the contents of phenomenal consciousness with the contents of sensory memory also fail, as discussed in the previous chapter.

8.2.2 Hybrid Theories

Hybrid theories attempt to take both reports and behaviours seriously but interpret them differently under different circumstances. Hybrid theories have been proposed by Tye (2009) and can be found in the many replies to Block (2007, see esp. Burge, Grush, Levine, Kouider et al. Papineau, Sergent and Rees, Naccache and Dehaene, Van Gulick). The general claim is that there are two types of visual processing with two associated types of phenomenology. Specific or item-based phenomenology stems from item-specific processing and is rich in detail but limited to the attended contents of consciousness. Generic or scene-level phenomenology stems from scene-level processing and contains conceptual information:

> We seem to have a full, richly detailed, phenomenal representation of the visual scene, though in fact what we have is, albeit full and clear . . . only actually detailed in some places and in some respects, and in other places and other respects it is clear but generic. (Grush 2007, p. 503)

This split between two types of phenomenology shifts the debate from one about the *amount* of content present in consciousness to one about the *type* of content present. However, as Block notes in his response to these proposals, having two types of phenomenology does not seem consistent with experience, particularly in the case of the Sperling paradigm. Letters are not experienced in a different way before the cue and then after it (pre- and post- focused attention). While splitting visual phenomenology into two types offers an easy way of explaining the contrasting reports and behaviours found in change and inattentional blindness, specifying the relationship between generic and specific phenomenology, and whether it actually matches the properties of generic and item-specific processing (discussed further below), is problematic.

8.2.3 Sensori-Motor Contingencies/Externalism

Sensori-motor models take reports of richness at face value, but understand perception to be temporally extended and intimately linked with action. These models locate the rich contents of consciousness in the world and our continuing interactions with it, rather than in a particular internal stage of perceptual processing (see e.g. Hurley 1998; O'Regan and Noë 2001). Thus we experience the world in a detailed way, where detail is delivered on demand by attending to it in the world. In discussing the apparent richness of the contents of consciousness, O'Regan and Noë state that "your *feeling* of the presence of all the detail consists precisely in your knowledge that you can access all this information by movements and inquiries" (O'Regan and Noë 2001, p. 960). However, it is difficult to explain the precise sense in which detail about potentially accessible information is 'felt', and how this 'felt' presence of potentially accessible information is characterised as actually and currently accessed information in reports (see also replies to O'Regan and Noë 2001).

8.2.4 Sparseness

Endorsed by a number of cognitive scientists (e.g. Dehaene et al. 2006; Mack and Rock 1998), the sparse view of conscious content typically accepts that reports are the only way to operationalise consciousness, but that they require careful interpretation. For example, it is largely agreed that subjects correctly report their experiences of attended items, but that reports of their experiences of unattended items or areas can provide a misleading description of the *current* contents of consciousness. Instead, it is suggested that subjects suffer from cognitive 'illusions' of thinking that they see more than they can concurrently access (Dehaene et al. 2006).

Supporters of the sparse view of conscious content suggest that while a whole scene is potentially visible via shifts of attention, very little of it is consciously perceived at one time. Behaviours, such as change detection, reflect this short-term sparse *actual* content, while scene-level reports reflect the longer-term, richer *potential* contents of consciousness. While this model provides a coherent account of perception of attended items, it can be argued that it fails to preserve standard intuitions about the breadth and relative stability of conscious content, as conscious content is limited to the few items present in working memory at any one time, that may change rapidly according to attentional and task demands.

8.3 Modelling Perception

The way that conscious content is described in the accounts above will be compared with a widely accepted model of perceptual processing, described here and illustrated below through a series of case studies. One important feature of the structure of perceptual processing is that different processing streams take input from the same areas or items and use it to fulfill different functions. As seen in the previous chapters, two of these streams comprise the ventral stream that processes perceptual input in an allocentric coding for object identification and goal selection, and the dorsal stream that processes perceptual input in an egocentric coding for online action guidance. At the very least, this shows that there is no single internal representation that mediates between visual information and reactions to stimuli.

Item-specific processing can also be contrasted with faster scene-level or gist processing, discussed in the previous chapter. In particular, gist processing can be heavily influenced both by expectation and by context-driven attentional demands. As gist processing is often used to direct attention towards specific items, the factors that affect gist processing also have an effect on the kinds of item-specific processing that subjects engage in. Expectations can strongly shape where attention is directed, and then affect not only whether a stimulus is reported, but what kind of stimulus it is identified as. Attentional demands can focus processing towards particular goals, but at the cost of decreased sensitivity to other non-relevant visual information. The way these factors affect reports and behaviours in practice is illustrated below in relation to partial report superiority, and change and inattentional blindness.

The ways in which expectation and attention affect visual processing are formally expressed in computational models that treat perception as a generative, inferential process (see e.g. Mumford 1992; Neisser 1967; Shams and Beierholm 2010; Yuille and Kersten 2006). In these models, perception is characterised as a process of hypothesis generation and testing, comparing higher-level hypotheses (expectations) to lower-level incoming information via feedback connections throughout the cortex. This changes both lower level stimulus processing and higher level hypotheses until a 'most likely' hypothesis emerges. There are of course variations in the details of specific computational models, including the degree to which different processing streams are independent of each other, but a coarse-grained description of these models is sufficient for the purposes of this chapter. This biologically plausible computational model of perception will be used below to explain the ways that different reports and behaviours can be generated based on the same informational input. This is then compared with the way that the contents of reports and behaviours are used to support different accounts of conscious content, in order to see if these accounts are scientifically viable.

Of crucial importance is the fact that the multi-stream, generative model of perception does not provide a single representation of visual information. This means that accounts of conscious content must provide some way of demarcating the set of conscious content among the contents of different processing streams. It will be shown that the contents exhibited in reports and behaviour can be identified with the contents of different processing streams that fulfill different sensori-motor and cognitive functions. However, it will be argued that no account succeeds in demarcating a set of content that is both phenomenally plausible, and that is consistent with the properties of perceptual processing. Following a discussion of how the Heuristic Identity Theory is applied to this case, it be argued that the concept of 'the contents of consciousness' is itself a problematic concept, and is unlikely to form a viable scientific category. The case studies used to motivate this argument are described below.

8.4 Partial Report Superiority, Change Blindness and Inattentional Blindness

As described in Chap. 7, well-practised subjects report seeing 'all the letters' in the partial report paradigm even though they are able to identify only three to four of them at once (Sperling 1960). It was argued in this chapter that reports of richness and letter identifications are generated by two very different processing streams. Reports of seeing the whole display are based on the contents of gist processing, and reports of letter identities are based on item-specific processing, both of which differ in their inputs, processing times, and function.

In particular, de Gardelle et al. (2009) note the different role of expectations ('priors') in these processing streams across situations in which there is more or less stimulus information. They argue that with low levels of stimulus information, gist hypotheses based on past experience of letter displays are used to generate reports about the whole display. Cases in which expectations do not match the display result in inaccurate scene level reports from subjects. They also suggest that while the heavy reliance on expectations in gist processing sometimes leads to inaccurate reports, gist hypotheses are typically very useful in generating appropriate reports and directing behaviours in the face of sparse information (e.g. from short stimulus duration times). The difference between gist and item-specific processing, and the significant role of expectation in gist processing when only poor evidence is available, needs to be accounted for in describing and isolating the contents of experience.

The role of expectation and attention in generating scene level reports, and in directing item-specific processing used to generate behaviours, can also be used to explain the phenomena of change and inattentional blindness. Change blindness consists in the failure to notice and report changes between otherwise identical scenes, either when presented across saccades or presented in quick succession (O'Regan et al. 1999; Rensink et al. 1997; Simons and Rensink 2005). Eye-tracking experiments have shown that when attention is directed at the changed item, the change is noticed and reported (e.g. Henderson and Hollingworth 1999). Similar to the Sperling paradigm, gist processing allows subjects to report that they have seen 'the whole scene' without being able to report any changes to it. Gist processing effectively ignores small variations in visual information, so changes are only noticed (reported) when attentionally directed item-specific processing is directed at the changed item across scenes.

Inattentional blindness, when subjects fail to notice salient features of a scene, highlights the role of attentional load in determining how much variance in the input to a gist hypothesis can be tolerated. In the standard basketball/gorilla example (Simons and Chabris 1999), subjects are asked to track the number of basketball passes a team wearing white t-shirts make as they move among a team wearing black t-shirts. During the video, a gorilla stands in the middle of the game and beats his chest, but this often goes unnoticed by subjects.

It seems that during the task, subjects form a gist hypothesis about the contents of unattended areas of their visual field (a number of moving, black, people-sized objects), and generate item-specific hypotheses for attended items (the basketball). Given the high attentional demands of the item-specific task, the gist hypothesis tolerates a wide range of variance in the input from non-attended areas. Subjects will therefore fail to notice the gorilla if it can be subsumed under the gist hypothesis. As gorillas are also black-people-sized-objects, the gist hypothesis goes unchallenged, so attention is not directed towards the gorilla and it is not noticed.

Even more striking instances of inattentional blindness, such as subjects failing to recognise massive changes in a conversational partner (e.g. Simons and Levin 1998), are likely based on similar expectation and attention based mechanisms. In these brief, real-time encounters, subjects are often in the middle of going somewhere, and are then asked to give directions to a (disguised) experimenter. Again, if subjects are distracted and clearly not expecting the conversational partner to change, then a hypothesis about the properties of the conversational partner may tolerate a reasonable degree of variance over time before being challenged and thus requiring attentional focus. Such a 'loose' hypothesis may be sufficient to capture all the information relevant for the subject at that time. Indeed, Simons (2000) suggests that inattentional blindness research may support the possibility that noticing *unexpected* objects may happen very rarely in the real world:

> This somewhat radical hypothesis would suggest that our intuitions about attentional capture reflect a metacognitive error: we do not realize the degree to which we are blind to unattended and unexpected stimuli and we mistakenly believe that important events will automatically draw our attention away from our current task or goals. (p. 154)

If gist hypotheses are heavily based on expectations, and item-specific processing is guided by an attentional set determined by the gist hypothesis, then it may indeed be very difficult to notice unexpected items (see e.g. Most et al. 2001, 2005). When unexpected items are noticed, item-specific reports will then conflict with earlier scene-level reports that are based on false expectations.

However, the picture is complicated further by evidence that some visual information about unreported changes and features are nonetheless stored in long term memory, and can be probed in forced-choice tasks (e.g. Hollingworth and Henderson 2002; Mitroff et al. 2004; Varakin and Levin 2006). While this difference may simply hark back to the earlier problem of which type of measure to use to assess consciousness, it also complicates the way in which conscious content should be demarcated. It is not clear if information that is not reported, but is stored in memory, forms part of the contents of consciousness. Perhaps only conscious content can enter long-term visual memory, or perhaps unconsciously processed contents can be stored there too. In this case, the contents of long-term memory, accessed via forced-choice paradigms, may provide no guide as to the (possible) contents of consciousness (parallel to the problems raised with interpretations of sensory memory in the previous chapter). In general, the kind of representations that can be part of conscious content (e.g. current vs. stored item-specific information, gist information) are difficult questions to answer. The ways in which different accounts of conscious content approach this problem, and problems with these approaches, are discussed below.

8.4.1 Gist and Item-Specific Information in Conscious Content

The brief accounts of the Sperling, change blindness and inattentional blindness paradigms above provide a number of problems for accounts of conscious content. As argued in the previous chapter, the rich view is particularly problematic as there is no evidence to suggest that a store of rich contents is necessary to generate scene-level reports of visual richness. Gist hypotheses are themselves sufficient to generate these kinds of reports. Furthermore, there are problems in identifying where such a store of rich content could be located within sensory processing. There is little evidence that such a store exists, and it is unnecessary to explain the reports and behaviours that the rich view appeals to.

However, sensori-motor accounts are almost designed to address the apparent problems raised by change and inattentional blindness, as these are paradigms in which time and action (including saccades) play a crucial role in whether or not changed or salient items are noticed and reported. Sensori-motor accounts suggest that item-specific information 'feels present' because we have sensori-motor knowledge of how to access external information. While dispensing with an internal store of rich information, the sensori-motor account preserves the notion of rich phenomenology.

However, if Simons' claim above is true, we may have much less veridical sensori-motor knowledge about unattended and unexpected items than the sensori-motor view supposes. That is, if we are often wrong about our degree of access to unattended and unexpected items, feelings of richness will be based on routinely mistaken sensori-motor contingencies. In this case, the world will not function as a rich external store of our 'felt' phenomenal content, as the 'felt' content may be quite different from what is present externally.

In part this is because gist hypotheses depend on previous sensori-motor activities, now instantiated as neural expectations. This means that gist hypotheses do not necessarily provide appropriate sensori-motor knowledge of how access information in a *current* environment. Further, as outlined in more detail in the previous chapter, gist hypotheses are sufficient to generate scene-level reports of richness. Particularly problematic for sensori-motor accounts, gist hypotheses are based only on past (not current) sensori-motor contingencies, and they are constituted entirely internally (for more on this kind of argument see replies to O'Regan and Noë 2001).

Hybrid views are different again but are also designed to address the worries raised by the three phenomena discussed above. According to hybrid accounts, phenomenal content is a mix of specific (attended) and generic (unattended) content. However, it is questionable where in the visual system a 'phenomenal representation' of a scene can be found that is constantly filled with a disjunctive mix of generic and specific phenomenology. Gist processing and item-specific processing occur concurrently (though proceed at different time-scales), and draw on different kinds of input from the same areas of a scene. Gist processing makes use of low-frequency visual information from all over a visual scene, while item-specific processing makes use of high-frequency visual information from limited areas. That is, the hypotheses formed in item-specific processing cover some of the same areas and the same items as gist processing. This means that there is no sense in which gist processing and item-specific processing provide a conveniently disjunctive representation of a scene. While being based on a real distinction between types of visual processing, the particular features of hybrid accounts have little biological plausibility; some areas of a scene are represented both in a generic and a specific sense. This problem of overlapping content is not recognised in hybrid accounts and it is not obvious how it could be resolved.

As the sparse view identifies conscious content with the current items present in working memory, it has no problems with its demarcation criteria, at least on scientific grounds. However, in doing so the sparse view could simply be described as an account about the contents of attentional focus and working memory. The sparse view also needs to provide an account of how the 'illusion' of being conscious of whole scenes comes about, particularly if gist hypotheses themselves are only the focus of attention for limited periods of time.

8.5 Demarcating Conscious Content

This brief outline of the problems in the way that gist and item-specific processing are dealt with in accounts of conscious content motivates a more general criticism of these accounts. This is that none of them offer a way of demarcating content that takes account of the properties of perceptual processing streams, and that is also consistent with standard properties attributed to phenomenal content. If an identity claim between conscious (visual) content and the content of visual processing is to be sustained, both sets of properties must be made consistent.

First, the demarcation problems for the hybrid and sparse accounts were given above. The hybrid account faces the problem that there is no evidence of a disjunctive representation of generic and specific information. No suggestion is made of how generic information is phenomenally present for only unattended areas, or how or where such a composite representation could emerge. While referring to real distinctions in processing streams, it is not clear if, or how, these different sets of content are unified into a disjunctive 'phenomenal representation'. Reasons must either be given for why there is no experience of overlapping sets of generic and specific phenomenology, why there are no obvious phenomenological shifts between generic and specific content, or how our conception of phenomenology must change in order to incorporate these notions of overlapping and shifting types of content.

Second, although the sparse view uses biologically plausible demarcation criteria for conscious content (they are simply demarcation criteria taken from cognitive science), they are plausibly best seen just as demarcation criteria for attentional focus and working memory. Little effort is made by supporters of the sparse view to account for the apparent 'illusion' of rich phenomenology, yet in doing so (i.e.

by referring to the ability to serially saccade and attend to many different specific items), it may simply collapse into a version of the sensori-motor view.

For example, Dehaene et al. (2006) suggests that the appearance of visual richness may depend on perceptions occurring actively over time, such that reports of visual richness "might arise because viewers know that they can, at will, orient attention to any location and obtain conscious information from it" (p. 210). In this case it must still be explained why perceptual consciousness does not appear, even on short time scales, as a series of snapshots consisting of the current contents of working memory. Alternatively, it must be explained how the apparent continuity of content itself is an illusion.

Demarcation criteria for the rich view suffer from a very different assumption, explored in more detail in the previous chapter. This is that contents of (phenomenal) consciousness exist at the same pre-categorical (or low categorical) or 'object' level. However, as explored earlier there is no place in the visual system where such content exists.

In contrast, sensori-motor accounts acknowledge the many ways in which subjects can interact with objects in their environment, and thus acknowledge the potential range of the contents of consciousness. Yet the way that sensori-motor accounts demarcate this content is again problematic. O'Regan and Noë (2001) state that:

> ... people are aware of what they see to the extent that they have control over that information for the purposes of guiding action and thought... visual awareness is a fact at the level of the *integrated behavior of the whole organism*... (O'Regan and Noë 2001, p. 969, added italics)

However, where to draw the line between content that does and does not contribute to the 'integrated behaviour of the whole organism' resembles the demarcation problems found in earlier chapters. For example, early global neuronal synchrony across perceptual areas that occurs *before* stimulus presentation plausibly contains contents (expectations about certain stimuli being presented) that guide later behaviours and thoughts (e.g. reports of gist, saccades and movements). However, it is difficult to see in what way the content of a pre-stimulus expectation could form part of the contents of consciousness. Also, while there is a great deal of sensori-motor content that drives behaviour, it is unclear if all sensori-motor expectations and motor planning can really contribute to the contents of consciousness. Finally, if visual awareness consists of those contents that contribute to behaviour, then it seems difficult to support a distinction between conscious and unconscious perception, as all behaviours directed by sensory information would count as providing conscious content. This problem could perhaps be avoided by shifting the burden to the concept of 'integrated behaviour', but again it is unclear what this really means.

This point follows from the general idea suggested elsewhere (e.g. rich and sparse views above, Lamme's mechanism of local recurrent processing discussed in Chap. 6), that demarcating conscious content and perceptual content in a consistent way is difficult. Thus Hardcastle (2001) concludes in a commentary on O'Regan and Noë that while learning about the visual system works can inform a theory of consciousness, that it does not in itself provide a theory of consciousness.

It is essential to recognise the full range of responses that can be made towards specific items (e.g. detection, identification), and the different types of processing found in the visual system (item-specific vs. gist processing), but doing so seems to come at a cost. This is that it becomes increasingly difficult to provide adequate demarcation criteria for the contents of consciousness. The lack of a short-term store of non-conceptual contents undermines the rich view entirely. There is no evidence of a disjunctive representation of generic and specific information as required for the hybrid account. Sensori-motor theories are likely to be adequate accounts of sensori-motor content but are too broad to function as theories of conscious content. Finally sparse views identify a very narrow set of content that also prevents them from isolating a phenomenally plausible set of content. Doing justice to the properties of perceptual processing yet capturing a set of content that satisfies assumptions about the phenomenal properties of consciousness is a problem for all accounts.

The discussions above have been short, and refer in some cases to well-trodden criticisms, but their main purpose is in highlighting the ways in which identity claims between the contents of consciousness and perceptual content can be attacked on both scientific and phenomenological grounds. What is of real interest is what is learnt from these criticisms, and how progress is made from advancing these identity claims. The act of suggesting identity claims, or cross-level theories, comes with its own heuristic value in promoting the revision of concepts across both levels, as mentioned above. The extent to which this heuristic is used, or can be used, in consciousness science is discussed below.

8.6 Identity Claims and Content-Matching

As has become clear over the earlier chapters and from the discussion above, there are a range of visual processes that support different kinds of responses and behaviours, including detection, categorisation and identification responses, 'free' and forced choice responses, first and second order confidence ratings, goal selection and online action guidance, scene-level and item-specific responses, and so on. Each of these processes takes different kinds of input, is sensitive to different factors and performs different functions. So far, it has been the job of cognitive science to identify these processes and functions. It is becoming increasing clear that there is no single representation of visual information from which responses and behaviours are generated, which can be easily identified with the contents of consciousness.

Instead, the ways in which we interact with the world are determined by a range of different, interacting, and sometimes mutually inconsistent hypotheses about a visual scene that are sensitive to expectation, attentional load, response bias, motor feedback, and temporal factors. Identifying a set of biologically plausible content that also satisfies assumptions about the phenomenal qualities of conscious content is a difficult task, as explored above. These assumptions can include the unity of conscious content, that content is represented in the same format (or same few formats), that it is greater than the contents of working memory but less than the entire contents of sensori-motor processing, and that it does not consist of overlapping or inconsistent content. Contemporary accounts of conscious content fail to identify a set of content that meets both these biological and phenomenological constraints.

This suggests that something is going wrong at either (or both) levels of description. If we are to continue to pursue the content-matching strategy, then concepts either at the philosophical/phenomenological or the scientific level (or both) need to be revised. McCauley and Bechtel's (2001) Heuristic Identity Theory suggests that making identity claims, examples of which can be found in the cross-level theories of consciousness explored above, serves to structure multi-level research. They do so by suggesting hypotheses, how to test them, and how to revise them when they fail. These revisions often include changes to the concepts found at both levels in the claim. In this way, concepts at all levels gradually become more sensitive to each other.

Craver (2007, see esp. pp. 258–261) also states similar ideas in his discussion of the constraints on inter-level integration. Accommodative constraints serve to modify taxonomies, mechanisms and theories across different research fields in both a top-down and a bottom-up way. Top-down accommodation places constraints on mechanisms and taxonomies in virtue of the way that the higher-level phenomena or mechanisms are described. Thus, by identifying instances of consciousness with attention and report, global workspace theories outline the sorts of content that can be conscious at any one time, and thus what kind of perceptual processing can provide this content. Bottom-up accommodation places constraints on the phenomena under investigation in virtue of discoveries about underlying mechanisms. This kind of accommodation was used in Chap. 6 to argue that mechanisms of cognitive and neural processes cannot support traditional definitions and taxonomies of consciousness. In the context of conscious content, it can also be used to argue that ways of taxonomising perceptual content do not map onto the way that conscious content is delineated.

Craver and McCauley and Bechtel state that identity claims and integrative techniques force revision at *all* levels involved. Models of perception are constantly revised due to top-down and bottom-up constraints given across psychology, neuroscience, computational neuroscience, and so on. Changes of this sort are expected in science, and eventually serve to provide a range of basic, and generally acknowledged, empirical facts and theoretical classifications.

However, the kind of revisions necessary to make models of perceptual processing more amenable to descriptions of conscious content would clearly conflict with deeply entrenched empirical findings and theories. Assumptions about conscious content are simply incompatible with many well-known and well-established facts about sensory and cognitive processing, such as the existence of multiple, overlapping and occasionally inconsistent contents of different processing streams. In this case there are strong reasons to ignore the specific top-down constraints of philosophical theories of conscious content onto scientific accounts of perceptual processing.

Any serious revision of theories of conscious content that would make them more amenable to identity claims with perceptual processing also seems unlikely, but for a very different reason. While models of perception are continually revised and corrected over time, through which they gain empirical and theoretical support, the basic conceptions we have about the contents of consciousness do not seem so amenable to change. While some are now moving away from rich accounts of content to hybrid or sensori-motor accounts (e.g. Tye 2009), most of the accounts that I am aware of still assume the unity, internal consistency, and non-overlapping nature of conscious content. The efforts that proponents of these accounts make in trying to locate such (rich) contents suggests that intuitions about conscious content die hard, despite their obvious inconsistency with scientific accounts of perception. The constraint of bottom-up accommodation of scientific theories onto philosophical or phenomenological ones therefore appears to function as a weak one. In this case, there is little possibility that sufficient revisions could be made that would allow (convincing) identity claims to be sustained between the contents of perceptual processing and the contents of consciousness.

However, one response to the problems outlined above is to simply identify consciousness with multiple, generative, temporally extended sets of content, and accept whatever revisions to the concept of 'consciousness' that are necessary as a result of this. Thus, according to Dennett's (1991) Multiple Drafts theory 'the contents of consciousness' simply refer to the current 'winning' hypothesis in the brain, i.e. whatever content is currently acted on or reported. There are always multiple drafts (hypotheses) in existence that may or may not be used in the future, and may have a greater or lesser direct effect on current behaviour, yet there is something like a central narrative that is the result of the (partial) integration of these drafts over time.

This account is completely in line with the model of perceptual processing outlined above. Dennett can explain the apparent (i.e. reported) richness of experience, the role of attention, the lack of apparent shifts in perceptual content due to the partial integration and revision of a central narrative, and all in a scientifically respectable way. On the face of it, this seems to be a prime example of higher-level concept revision in light of empirical findings. However, like sensori-motor theories of conscious content described above, it is not clear what is gained by calling the Multiple Drafts theory a theory of *consciousness*. The Multiple Drafts Theory is a theory about multi-stream, temporally extended information processing and the generation of behaviours, but it is plausibly best described in these terms. Indeed, as Korb's review of *Consciousness Explained* (1993) states, "I believe that the central thesis will be relatively uncontentious for most cognitive scientists, but that its use as a cleaning solvent for messy puzzles will be viewed less happily in most quarters" (Sect. 1.2).

This is partly because in attempting to answer these 'messy puzzles' as puzzles about *consciousness*, Dennett is continuing to use of the concept of consciousness while trying to give it a very different meaning. Instead, many critics of Dennett think that he just doesn't 'get' the questions surrounding consciousness, so that *Consciousness Explained* certainly does *not* explain consciousness. Far from being an acceptable solution to the problems found in the content-matching strategy, the rejection of phenomenological assumptions and traditional questions about consciousness, combined with the preservation of the term 'consciousness', strikes many as unacceptable.

While generally supportive of Dennett's strategy, I suggest that the revisions required in concepts of consciousness to make them scientifically viable (which may be more extreme than suggested in Dennett's account) are so deep and (often) counter-intuitive, that it is plausibly *not* the best way to deal with failures in content-matching strategies. The revisions that are necessary to make the concept of 'the contents of consciousness' consistent with scientifically plausible models of perceptual processing are so extreme that the concept would no longer be recognisable.

This seems to largely account for the criticisms of Dennett's Multiple Drafts account (and possibly sensori-motor and sparse theories too); it just does not *look* like an account of consciousness, and states that many of the standard intuitions and puzzles about consciousness are nonsensical. While other instances of concept revision can incorporate massive changes in the core properties attributed to the concept (e.g. electrons), such massive changes seem to be unacceptable in the concept of consciousness. In this case, the computational model of perception described above is *not* offered as a model of conscious content, but simply as a model of perception, against which standard phenomenological properties can be compared.

Therefore, accounts of conscious content either get the phenomenology 'right' but are not consistent with the details of visual processing, or they offer a scientific theory of perceptual processing that disregards standard phenomenological properties of conscious content. There seems to be no available middle ground between accounts of conscious content and accounts of perceptual processing because they embody such very different properties. As noted above, accounts of conscious content that ignore standard intuitions about consciousness, or explain them away as illusory, are rarely seen as acceptable accounts. So, radical change in how we describe the contents of consciousness is required if the content-matching strategy is to succeed, but within the community of consciousness researchers, this is not an acceptable move.

This strongly suggests that the research heuristic associated with advancing identity claims does not, and cannot function, within the content-matching strategy in consciousness science. Not only are accounts of (visual) conscious content and the contents of the perceptual system incompatible, but this is a state that is actually sustained and promoted in contemporary consciousness science. If the concepts used by a research community do not allow that concept to be used in standard research practices, this is reason enough to question the place of that concept in scientific research.

8.7 Conclusion

The failure of the heuristic associated with cross-level research in the case of content-matching provides further support for the claim concepts of consciousness impede standard scientific research practice, so are not viable scientific concepts. The content-matching strategy ignores the massive differences between the properties of perceptual processing and the properties of conscious content. Crucially, the content-matching strategy also ignores the heuristic role that advancing identity claims usually fulfills. Concepts at both sides of the identity claim go unrevised in light of each other. Dennett's solution, in rejecting the properties typically attributed to consciousness, is often seen as unsatisfactory in the same way that sensori-motor and sparse theories are unsatisfactory; they simply leave questions about *consciousness* unanswered. This sociological feature of the research community is an important one, as it underlines the degree to which concepts of consciousness are unlikely to change, and thus how accounts of conscious content will continue to be incompatible with empirical research.

Taken together with similar claims about the inability to find behavioural or neurophysiological measures of consciousness, or a mechanism for consciousness, the failure of the content-matching strategy suggests that the project of pursuing a science of consciousness is simply untenable. These claims are based purely on assessments of how well consciousness science currently uses standard scientific methods, such as dissociation, integration, demarcation, and inter-level identification (matching). Its failures to use these methods appropriately, and the reasons why these methods cannot be used appropriately in relation to 'consciousness' suggest that, from a matter of scientific practice, 'consciousness' is not a viable scientific concept. Concepts of consciousness should therefore be eliminated from science as a viable target of research.

In rejecting 'consciousness' as a scientifically useful term altogether, we can instead identify an appropriate set of concepts and research questions to investigate the diverse capacities and behaviours explored throughout the book. Answering questions about these capacities and behaviours can be done in a way that respects scientific methodology, but leaves out the unanswerable question which ones *really* tell us about consciousness. A detailed argument against the viability of 'consciousness' as a useful scientific term is provided in the next chapter.

References

Block, N. (2007). Consciousness, accessibility, and the mesh between psychology and neuroscience. *Behavioral and Brain Sciences, 30*, 481–548. (inc. responses).

Craver, C. F. (2007). *Explaining the brain: Mechanisms and the mosaic unity of neuroscience*. Oxford: Clarendon Press.

Gardelle, V. de, Sackur, J., & Kouider, S. (2009). Perceptual illusions in brief visual presentations. *Consciousness and Cognition, 18*, 569–577.

Dehaene, S., Changeux, J. P., Naccache, L., Sackur, J., & Sergent, C. (2006). Conscious, preconscious, and subliminal processing: A testable taxonomy. *Trends in Cognitive Sciences, 100*, 204–211.

Dennett, D. C. (1991). *Consciousness explained*. London: Penguin.

Grush, R. (2007). A plug for generic phenomenology. *Behavioral and Brain Sciences, 30*, 504–505.

Hardcastle, V. (2001). Visual perception is not visual awareness. *Behavioral and Brain Sciences, 24*, 985.

Henderson, J. M., & Hollingworth, A. (1999). The role of fixation position in detecting scene chances across saccades. *Psychological Science, 10*, 438–443.

Hollingworth, A., & Henderson, J. M. (2002). Accurate visual memory for previously attended objects in natural scenes. *Journal of Experimental Psychology: Human Perception and Performance, 28*, 113–136.
Hurley, S. (1998). *Consciousness in action*. Cambridge: Harvard University Press.
Korb, K. (1993). Stage effects in the Cartesian theater: Review of consciousness explained. *Psyche, 1*(4).
Lamme, V. A. F. (2006). Towards a true neural stance on consciousness. *Trends in Cognitive Sciences, 10*, 494–501.
Mack, A., & Rock, I. (1998). *Inattention blindness*. Cambridge: MIT Press.
McCauley, R. N., & Bechtel, W. (2001). Explanatory pluralism and heuristic identity theory. *Theory and Psychology, 11*, 736–760.
Mitroff, S. R., Simons, D. J., & Levin, D. T. (2004). Nothing compares 2 views: Change blindness can occur despite preserved access to the changed information. *Perception and Psychophysics, 66*, 1268–1281.
Most, S. B., Simons, D. J., Scholl, B. J., Jiminez, R., Clifford, E., & Chabris, C. F. (2001). How not to be seen: The contribution of similiarity ad selective ignoring to sustained inattentional blindness. *Psychological Science, 12*, 9–17.
Most, S. B., Scholl, B. J., Clifford, E. R., & Simons, D. J. (2005). What you see is what you set: Sustained inattentional blindness and the capture of awareness. *Psychological Review, 112*, 217–242.
Mumford, D. (1992). On the computational architecture of the neocortex. II. The role of cortico-cortical loops. *Biology and Cybernetics, 66*, 241–251.
Neisser, U. (1967). *Cognitive psychology*. New York: Appleton-Century-Crofts.
O'Regan, J. K., & Noë, A. (2001). A sensorimotor account of vision and visual consciousness. *Behavioral and Brain Sciences, 24*, 939–1031.
O'Regan, J. K., Rensink, R. A., & Clark, J. J. (1999). Change-blindness as a result of 'mudsplashes'. *Nature, 398*, 34.
Rensink, R. A., O'Regan, J. K., & Clark, J. J. (1997). To see or not to see: The need for attention to perceive changes in scenes. *Psychological Science, 8*, 368–373.
Shams, L., & Beierholm, U. R. (2010). Causal inference in perception. *Trends in Cognitive Sciences, 14*, 425–432.
Simons, D. J. (2000). Attentional capture and inattentional blindness. *Trends in Cognitive Sciences, 4*, 147–155.
Simons, D. J., & Chabris, C. (1999). Gorillas in our midst: Sustained inattentional blindness for dynamic events. *Perception, 28*, 1059–1074.
Simons, D. J., & Levin, D. T. (1998). Failure to detect changes to people during a real-world interaction. *Psychonomic Bulletin and Review, 5*, 644–649.
Simons, D. J., & Rensink, R. A. (2005). Change blindness: Past, present, and future. *Trends in Cognitive Sciences, 9*, 16–20.
Sperling, G. (1960). The information available in brief visual presentations. *Psychological Monographs: General and Applied, 74*, 1–29.
Tye, M. (2009). *Consciousness revisited: Materialism without phenomenal concepts*. Cambridge: MIT Press.
Varakin, D. A., & Levin, D. T (2006). Change blindness and visual memory: Visual representations get rich and act poor. *British Journal of Psychology, 97*, 51–77.
Yuille, A., & Kersten, D. (2006). Vision as Bayesian inference: Analysis by synthesis? *Trends in Cognitive Sciences, 10*, 301–308.

Chapter 9
Scientific Eliminativism: Why There Can Be No Science of Consciousness

9.1 Introduction

The previous chapters have outlined several sets of methodological problems in consciousness science. One concerns the problems involved in trying to establish a measure of consciousness. A range of subjective and objective behavioural measures and related neurophysiological measures were assessed in Chap. 2–5. It was argued that while these measures are appropriate measures for stimulus sensitivity, report, second order confidence ratings, and so on, there seem to be no principled way of identifying any of these as measures of consciousness. Given the current literature, there is also little possibility that a pragmatic agreement on an appropriate measure will arise. Following from this, another set of problems concerns which (if any) distinct category of phenomena are referred to by concepts of consciousness, discussed in Chap. 5–6. Here, it was suggested that the proper application of scientific methods to consciousness suggests that there will be no convergent measures and no distinct neural mechanism for consciousness, only measures of a variety of different cognitive and neural phenomena. The final methodological problem discussed in the previous two chapters is the significant mismatch between the phenomenal properties attributed to conscious content and structure of perceptual processing, and the lack of any way of dealing with this mismatch.

In order to show that 'consciousness' is not a viable target of scientific research, it has been argued that standard scientific methodologies, such as dissociation methods, integrative techniques, methods to demarcate mechanisms, and the proposal of cross-level identity claims, cannot be successfully applied to concepts of consciousness. Aside from failing in particular instances these methods also fail to fulfill their crucial heuristic role in the practice of science, for example by providing more clear and precise definitions of the phenomena under investigation, generating new research questions, and revising concepts across different levels of description.

It could be argued that these problems are the result of consciousness science currently being an immature science, so not yet possessing appropriate experimental methods or theoretical frameworks, or that they only apply to the specific cases considered. However, arguments have been offered to show that these are chronic problems resulting from concepts of consciousness themselves. This final chapter

E. Irvine, *Consciousness as a Scientific Concept*, Studies in Brain and Mind 5,
DOI 10.1007/978-94-007-5173-6_9, © Springer Science+Business Media Dordrecht 2013

explores the general factors used to support scientific eliminativist claims, in order to argue that they too apply to 'consciousness'.

9.2 Identifying Target Phenomena

Looking at the ways in which science can misidentify target phenomena provides a rough first pass at examining eliminativist claims about consciousness. If a target phenomenon is not identified appropriately, i.e. it can be shown that the concept used to refer to the phenomenon is incoherent, or seems to refer to a phenomenon that does not exist, then that concept can be safely and productively eliminated from science. In particular, the research questions posed in earlier stages of a science can be later seen as ill-posed, as the target phenomena referred to in these questions may be incorrectly identified and described. Craver (2007) identifies three ways in which an explanandum (target) phenomenon can be wrongly described: underspecification, taxonomic errors, or misidentification (pp. 122–128). Only misidentification can be used to support outright eliminativism. It will be argued that the science of consciousness not only makes the two less serious errors, from which a science can still progress, but that it is done so in such a way as to provide evidence for the error of misidentification. Further support for this claim is then provided in subsequent sections.

9.2.1 Underspecification

One possible error in identifying a target phenomenon is that it is underspecified. By failing to describe the multifaceted character of the phenomenon, conditions necessary for its occurrence, modulation or inhibition, how it behaves under a wide variety of conditions, and what its by-products are, a range of problems ensue. Given a lack of clarification about what the phenomenon is and how to identify it, it can be difficult to agree on a standard working definition of it, leaving it an open question whether different research groups could in fact be investigating very different phenomena. In this case, conflicting theories and mechanisms for the 'same' phenomenon can be put forward by different research groups. This kind of error was clearly seen in Sullivan's (2009) description of attempts to investigate LTP in neuroscience, in Chap. 5.

It was argued that this problem is also present in consciousness science. Although it is deemed obvious what consciousness is, earlier chapters (particularly Chap. 2, 3 and 5) show that there are a wide range of possible operationalisations and measures of consciousness. The standard taxonomies that typically use attention or working memory to separate phenomenal or pre-conscious processing from access or conscious processing offer some degree of clarification. However, different task types and different ways of gathering reports can be used to identify very different mechanisms, even for the same 'type' of consciousness. In particular, it was argued in Chap. 6 that 'reportability', a concept often used to operationalise and define consciousness, does not refer to a single phenomenon or mechanism, but a range of task-specific phenomena and a range of task-specific mechanisms. Consciousness is deeply underspecified, and far more than is currently recognized. This in itself leads to many of the problems and debates found in consciousness science.

9.2.2 *Taxonomic Errors*

The second possible error in the identification of a target phenomenon is that a set of phenomena are not appropriately categorised, either as a result of 'lumping' together many different phenomena, or 'splitting' a group of similar phenomena. As evidenced by Bechtel (2008, esp. pp. 49–88), memory was lumped as a single phenomenon, and assumed to have a single set of mechanisms for encoding, storing and retrieving information. However, it is increasingly clear that the mechanisms for memory overlap with other abilities, such as perception, and that different types of memory do not appear to be based on the same mechanisms. Splitting phenomena can also lead to fractured research on a group of phenomena that all turn out to be generated through the same mechanism.

Again, errors in taxonomisation are also present in consciousness science. Stemming from the problem of underspecification, current taxonomies of consciousness are too broad and lump together diverse sets of behaviours and abilities. This is clearly evidenced in the availability of a huge and often conflicting range of measures and mechanisms even for the same type of consciousness. Further, as argued throughout earlier chapters, current taxonomies fail to map onto the real distinctions found in cognitive and neural processing. For example, there are methodologically valid distinctions to be made between sensitivity and response bias, and between first and second-order confidence ratings, but not between conscious and unconscious processing, or between phenomenal or access consciousness (Chap. 2–3). Likewise there is a methodologically viable distinction between mechanisms of sensory processing and decision-making, but not between mechanisms that produce content, and mechanisms that make that content conscious (Chap. 6). Indeed, it was claimed at the end of Chap. 6 chapter that a major problem with consciousness science is that it lumps together many different phenomena by stipulation, and ignores scientifically valid distinctions that have previously been drawn between them.

9.2.3 *Misidentification*

Finally, the most serious error is the misidentification of a target phenomenon, in which the phenomenon simply does not exist. Standard cases of misidentification in the history of science are phlogiston and aether, but Craver also notes the study of animal spirits as a way of explaining the actions of nerves (Craver 2007, p. 123). The way in which the two errors of underspecification and mis-taxonomisation have occurred in consciousness science also suggests that 'consciousness' is not a viable target phenomenon. However, the claim is not that consciousness does not exist (whatever that might mean), but that it does not constitute a viable scientific concept. An examination of the range of operationalisations, measures and mechanisms of consciousness show that there is nothing in common between them all, and that instead they refer to a wide range of different phenomena and mechanisms already described in cognitive science. While it is possible to label these different phenomena as subtypes of consciousness, the practical and explanatory benefits of having two sets of labels of the same phenomena, one comprised on the vocabulary of perceptual and cognitive abilities, and one with the word 'consciousness' added, is of little practical or explanatory benefit.

This section provides only a short and simple evaluation of whether 'consciousness' and its subtypes are appropriate target phenomena for scientific research is suggestive. Clearly, a more detailed assessment of criteria for the elimination of concepts from science is necessary. Many of these criteria are discussed in earlier chapters, though often briefly and only implicitly. Based on methodological problems, the identification of scientific kinds, and epistemological and pragmatic factors, these criteria are discussed below to provide more detailed support for the claim that 'consciousness' is a misidentified phenomenon and should be eliminated from scientific vocabulary.

9.3 Scientific Eliminativism

Although there are several strategies to use when arguing that a concept should be eliminated from science, they are all elaborations of the basic idea that concepts should only be preserved so long as they promote the typical goals of science (e.g. to describe and predict phenomena), and that they preserve standard scientific practices. If concepts block progressive research then they should be eliminated. This final argument for the elimination of 'consciousness' in scientific research splits into four parts: the first on the valid or invalid application of scientific methods, the second on the implications this has on identifying scientific kinds, the third on broader epistemological factors in promoting viable scientific research programs, and finally further pragmatic factors in the practice of successful science. The criteria identified in these sections support the claim that eliminativism about concepts of consciousness is warranted on many fronts.

9.3.1 The Argument from Scientific Method

One of the main aims of this book has been to show that standard scientific methods are not, and cannot, be successfully applied to concepts of consciousness. It could be argued that we may simply be missing appropriate methods, and that they will be developed through further research. However, the fact that very basic methods, like dissociation, demarcation and integrative methods, fail to be applied successfully, and that methods designed for consciousness science such as the content-matching strategy also fail, suggests that there is a serious problem in consciousness science.

As argued in Chap. 3 and 4, dissociation methodology encounters problems both in its application in establishing particular behavioural measures, such as d', as measures of consciousness, and in its more general application and heuristic role in the science of consciousness. Dissociations are typically used as a way of investigating the structure of sensory or cognitive systems, by progressively refining the descriptions of phenomena under investigation, and testing and validating taxonomies and research questions. For example, dissociations in the ventral and dorsal streams of perceptual processing led to the hypothesis of two separate streams for the 'what' and 'where' of visual objects (Ungerleider and Mishkin 1982). More recent experimental work, again using dissociation methods, provides a more precise definition of the function of these pathways (Milner and Goodale 2008), and shows the degree of communication between them (Schenk and McIntosh 2010). By promoting a virtuous circle of mutual refinement between the interpretive frameworks used to classify dissociated phenomena, experimental paradigms used to test the frameworks, and the generation of future research questions, dissociations allow more and more precise definitions of dissociated phenomena to be made.

However, dissociations in the literature on consciousness are not used in this way. Chapter 3 showed how apparent dissociations between conscious and unconscious perception were better described as dissociations between sensitivity and context-sensitive decision-making and report. It was shown also shown how attempts to establish a distinction between conscious from unconscious perception to get around the debate between objective and subjective measures ultimately collapses into this debate anyway. There are no principled methodological solutions for this problem, and seemingly little chance of a pragmatic, community-wide agreement on how to proceed.

It was also argued in Chap. 4 that the virtuous circle between the development of experimental paradigms, and the frameworks used to interpret the results of these paradigms, cannot be found in consciousness science. The details of experimental manipulations are not used to inform interpretive frameworks; categories of conscious and unconscious perception are simply attached to dissociations. Proposed dissociations between conscious and unconscious perception are not further tested, or categories revised in light of new findings; instead researchers seem to be searching for the 'right' dissociation for the labels of conscious and unconscious perception to attach to, that best coincides with their pre-theoretical assumptions. That is, instead of developing new theories given experimental findings, we strain to fit our intuitive conceptions about consciousness onto existing categories and distinctions.

The integrative approach used to establish measures and taxonomies of consciousness discussed in Chap. 5 is also problematic. While integrative methods are usually productive ones (cognitive science is a field that integrates research from many disciplines), the approach of integrating and thus converging on measures of consciousness fails for several reasons. First, using Sullivan's (2009) approach of examining the diversity of experimental practices, it was shown that the methods of operationalising consciousness are so diverse that researchers across different lab groups are plausibly investigating very different phenomena. This would also explain the divergence in their experimental results. Further, the convergence that *can* be observed across measures of consciousness occurs within groups of measures that index the same behavioural operationalisation of consciousness, and so is entirely predictable and uninformative given background knowledge of these behaviours.

In attempting to integrate a wider range of measures and experimental paradigms, no convergence or set of common properties can be found. The failure to find global convergence across a range of measures also shows that 'consciousness' cannot function as a scientifically useful umbrella term. Its diverse referents often have very little in common, ensuring that they cannot be identified with a unique set of common properties or a common function. In this case 'consciousness' fails to function as a viable target of research.

The problems of failing to clearly identify a target phenomenon are also echoed in the discussion of the search for a mechanism of consciousness (Chap. 6). Demarcating mechanisms demands an exploration of the causal structure that lies behind the phenomenon to be explained, such that background conditions can be separated from constitutive components. Using criteria based on the notion of mutual manipulability (Craver 2007), mechanisms are isolated such that they include components whose action significantly affects the target phenomenon, and such that changes in the target phenomenon also significantly affect the components. First, it was argued that criteria of constitutive relevance show that there is no mechanism for reportability, but a range of task-specific mechanisms for task-specific phenomena. These mechanisms are also more inclusive than proposed by Global Neuronal Workspace theories. Further, demarcation criteria show that the distinction between the contents of consciousness, and consciousness of that content, is simply a confused way of referring to the distinction between sensory processing and decision-making. Second, criteria of internal consistency undermine the demarcation of Lamme's mechanism of local recurrent processing for phenomenal consciousness. Demarcation criteria are not appropriately applied in consciousness science, and when properly applied they identify quite different mechanisms for quite different phenomena.

Finally, the content-matching strategy was the subject of Chap. 7–8. It was argued that one popular example of this approach, mapping the contents of phenomenal consciousness with the contents of sensory memory, fails from a misunderstanding of the structure of the visual system. It was then argued that the strategy fails more generally as it is impossible, from an empirically informed model of perceptual and cognitive systems, to identify a stage of processing with 'the contents of consciousness', as standardly construed. Given the fundamental incompatibility of the structure of perceptual and cognitive processes with the properties of 'the contents of consciousness', identity claims between the two simply cannot be sustained.

Drawing on McCauley and Bechtel's Heuristic Identity Theory (2001) of the role of identity claims research practice, it was argued that the radical reconceptualisation of either of these two levels of description is fundamentally unlikely, though this would be necessary to preserve an identity claim between them. The failure of the heuristic associated with identity claims to provide any cross-level hypothesis that satisfies both philosophical/phenomenal and biological constraints shows that no progress can be made using the content-matching strategy. This again underscores the claim that the science of consciousness does not, and cannot, function as a science.

These arguments show that a series of standard and indispensable methods used in science, particularly important in the history of the cognitive sciences, simply do not work in the science of consciousness. The method of content-matching, specific to consciousness research, also fails as a valid scientific method. This is not because the science of consciousness is a new field, making only tentative claims about measures and mechanisms of a phenomenon, but is due to a chronic problem in its assumptions, concepts and research questions. The failure of the science of consciousness to satisfy standard methodological norms strongly suggests that 'consciousness' and related concepts can be safely and productively eliminated from the scientific domain.

9.3.2 The Argument from Scientific Kinds

Boyd (1991) and Kornblith (1993) argue that scientific kinds in science can be identified by looking for clusters of commonly co-occurring properties that are generated by a common mechanism (homeostatic property clusters). Scientific kinds are just those things that science can identify by looking for convergent properties, and applying demarcation criteria to mechanisms. This puts the definition and identification of scientific kinds firmly in the realm of scientific practice, and in turn means that scientific kinds are just those things that it is possible to make generalisations and predictions about. Arguing that 'consciousness' fails to pick out a scientific kind (or set of kinds) provides further support for an eliminativist claim about consciousness. Chapter. 5 and 6 provided these arguments by identifying the clusters of properties that can be found across consciousness science, and by identifying the mechanisms that can and cannot be found for phenomena used to operationalise consciousness.

Chapter 5 offered arguments as to why clusters of properties could be found for kinds of sensory and cognitive processes, but not for 'consciousness' or any of its sub-types. It was argued that the clusters of properties found should be described as the product of the precise behavioural operationalisation of consciousness that is used in an experimental paradigm. These property clusters are relative to the particular sensory or cognitive processes that a task assesses, and so are clusters of properties related to sensory processing, attention, decision-making, and so on. In labelling property clusters as kinds of consciousness, this obscures how and why they are differentiated, and makes it very difficult to make predictions and generalisations about how the phenomenon would change under a range of experimental manipulations. Further, it was argued that there are no unique common properties found across all measures of consciousness. Identifying property clusters cannot help in identifying consciousness as a scientific kind, or a set of related scientific kinds. (The way that pragmatic factors in science give further support to this claim are discussed further below).

In Chap. 6 the mechanisms of consciousness proposed in Global Neural Workspace Theory and the Neural Stance were examined to assess if they were well demarcated, and whether they in fact gave rise to the phenomena they were directed at. It was shown that in contrast to the assumption found in GNWT, 'reportability' does not pick out a single, coherent phenomenon but refers to a range of task-specific phenomena that are realised by a broad range of task-specific mechanisms. Importantly, mechanisms for reportability cannot be seen as comprising a set of similar mechanisms, and thus themselves forming a scientific kind. The fact that it is very difficult to make predictions or generalisations about whether or not a subject will report a stimulus *in general* is the reason that cognitive science investigates the scientific kinds picked out by different types of sensory processing, decision-making, and attention. It was also argued that the mechanism identified in the Neural Stance was not demarcated in a consistent way. Further, it was unlikely that it could be identified as the mechanism for consciousness due to the incompatibility between the properties attributed to neural processing and those attributed to phenomenal consciousness.

From these two examples it was suggested that other top-down and bottom-up approaches to identifying a mechanism/mechanisms for consciousness, and therefore identifying consciousness as a (set of) scientific kinds, will also fail. As suggested by Lamme, top-down approaches seem incapable of identifying coherent high-level target phenomenon that are distinct from those already investigated in the cognitive sciences (e.g. attention, working memory), and are always tied to problematic behavioural measures. Bottom-up approaches face the problem of matching properties of neural processes with incompatible properties of consciousness. In this case, the scientific kinds that can be identified using Boyd's HPC account are not kinds of consciousness, but kinds already identified in the cognitive neurosciences. Again, 'consciousness' is not a viable target for scientific investigation, and instead promotes methodologically flawed research programs.

9.3.3 Epistemological Factors

However, even when there are reasons to believe that concepts do not pick out scientific kinds, they can still play a useful role in scientific practice, and therefore should not be eliminated. Brigandt's (2003) discussion of 'investigative kind concepts' suggests that epistemological factors can play a role in deciding whether or not a concept should be eliminated from science. Brigandt's argument stems from a debate in biology about whether the concept 'species' should be eliminated. Ereshefksy (1992, 1998) has argued that 'species' splits into three distinct kinds of ecospecies, biospecies and phylospecies, each of which pick out different groups of organisms, and thus that the overarching concept 'species ' can be eliminated. However, Brigandt argues that the concept 'species' can still play a valuable role in guiding research, and his account can be extended to other cases. He argues that if a concept can be used in coherent research questions and is used to guide research on the mechanisms underlying the phenomena in question, then the concept can still function as an 'investigative kind' (INK), and should not be eliminated.

Brigandt describes the two conditions that must obtain for an INK concept to be eliminated. First, a concept can be eliminated if it cannot figure in theoretical generalisations and explanations across a range of theoretical contexts or across a research field. Second, if there are empirical findings that undermine the theoretical motivation for using the term, which lead to different theoretical needs, then this also contributes to the need to eliminate it.

Machery (2009) and Griffiths (1997) have used similar criteria to argue that the concepts 'concept' and 'emotion' should be eliminated from science. For example, Machery argues that current research suggests that the concept 'concept' actually refers to three different kinds of information (exemplars, prototypes, and causal information), which can be dissociated and can be used to perform different tasks. These more specific kinds of 'concept' do not support generalisations about all concepts, and do not share the features of the original term. Similarly, Griffiths states that 'emotion' refers to "... affect programs, domain-specific biases in motivation, socially sustained pretences, and other more specific categories of psychological state and process that have been identified or hypothesized in the varied literature that sets out to address human emotion" (Griffiths 2004, p. 902). Generalisations cannot be made about all these kinds, and many have little to do with the original concept of 'emotion'. Keeping the concepts of 'concept' and 'emotion' therefore suggests that there is far more similarity in the group of kinds they refer to than actually exists, and promote useless discussions on the 'core' features of the concepts. As a consequence, 'concept' and 'emotion' do not refer to scientific kinds, but they are *also* no longer useful in scientific research as they do not promote fruitful avenues of research.

The case appears even more clearly in consciousness science. Researchers are finding that it is very difficult to generalise across measures and mechanisms of consciousness, even when the same kind of consciousness is under investigation (typically access consciousness/reportability). Chapter 2 showed that the presence of variable response bias means that the contents of subjective reports (both first and second-order) change across contexts and tasks. Chapter 3–4 showed that the significant problems that exist in identifying appropriate qualitative differences and dissociations in behaviour ensure that a range of very different phenomena are used to measure consciousness. This alone means that generalisations made across consciousness, even across the same 'type' of consciousness, are hard to come by. Chapter 5 and 6 elaborated on this to show where convergence can and cannot occur across and between behavioural and neurophysiological measures. This evidence can be used to outline the range of very different sensory and cognitive phenomena referred to as instances of consciousness. The stark differences between these phenomena, recognised elsewhere in cognitive science, prevent these phenomena from supporting generalisations.

The second point, whether the groups of phenomena that predictions and generalisations can be made about (i.e. the real scientific kinds), have much in common with the original concept of 'consciousness', is yet more obvious. The contrast between 'consciousness' and the examples mentioned above is illustrative of this. The concept 'concept' splits into prototype, examplar, and causal sets of knowledge, so at least part of the original meaning of 'concept', as bodies of knowledge that can be drawn on to categorise information, is preserved. The concept 'species' can be split into ecospecies, phylospecies and biospecies (see Ereshefksy 1992, 1998). However, as argued elsewhere (e.g. Wilson 1999), these apparently different definition of 'species' may have a lot in common due to environmental, evolutionary and developmental constraints, and at least identify sets of (often overlapping) similar organisms, thus preserving at least some of the original meaning of 'species'. For 'emotion', affect programs, motivation biases and so on, while clearly related to 'emotion', are less identifiable with the personal level of description usually invoked by the concept.

The case with consciousness is again much stronger. As argued in many of the previous chapters, consciousness does not split into an easily manageable number of distinct types of consciousness, all sharing a group of related properties that are easily associated with the personal level phenomenon of consciousness. Attempts to provide behavioural measures (Chap. 2 and 3), neurophysiological measures (Chap. 5), neural mechanisms (Chap. 6), and neural correlates (Chap. 7 and 8) of phenomenal consciousness all fail because differing operationalisations split 'consciousness' into one of the many varied and fine-grained scientific kinds. These include different types of attention, different streams of information processing, decision-making, and so on. These kinds seem to have very little in common with the original concept of (phenomenal) consciousness. This is of course why critics of consciousness science claim that experience can never be given a scientific explanation (Chalmers 1995; Levine 1983), or that many operationalisations of consciousness have actually got nothing to do with phenomenal experience (Block 2007; Lamme 2006). That is, while the effects of masking or attention on performance at different tasks can tell us how we process and respond to information, it is often difficult to see how this kind of research can tell us about consciousness.

Upon empirical investigation, the concept of consciousness simply shatters in a way totally unlike the cases described above. In practice, 'consciousness' is used to refer to a vast number of very different phenomena that often have little in common with each other, or with philosophical and phenomenological descriptions of consciousness. The failure of 'consciousness' to figure in generalisations, or to refer to the kind of phenomena that are actually being investigated, shows that it fails even as an Investigative Kind Concept. Again, this suggests that the concept 'consciousness' should be eliminated from science.

9.3.4 Pragmatic Factors

Building on these arguments, further pragmatic criteria for eliminating concepts from science can be considered. While pragmatic criteria may be seen as rather weak criteria to use, they are incredibly powerful means of organising a science and determining how it is carried out. In particular, these criteria allow concepts to be identified as either useful or harmful in promoting good scientific practice. Whether a concept is entrenched in a discipline, whether it promotes stability, continuity and generality, and whether the concept is ambiguous, all are ways of identifying 'problem' concepts. These pragmatic factors largely reflect the role concepts play in communication between scientists (crucial for debates about conflicting theories), and in charting the progress of a field of study. For ease of exposition, the factors discussed below are posed as criteria for *keeping* a concept, rather than elimination.

9.3.4.1 The Concept Is Entrenched

If a concept is methodologically problematic, yet entrenched and difficult to get rid of, then it might not be worth eliminating it. Concepts can be entrenched in the teaching of a science, and in the way that theories are stated. While Ereshefsky has argued that 'species' should be eliminated from science (1992, 1998), in a more recent paper he accepts that it is so central to biological discussions that it would not be beneficial to eliminate it (2009).

The science of consciousness, as opposed to the fields of biology that make use of the 'species' concept, has not been around for very long, so is not well entrenched. Consciousness was studied by a few psychologists and psychophysicists starting at the end of the nineteenth century, but behaviourism largely prevented much research being carried out on this phenomenon through the first half of the twentieth century. However, in the 1980s, neuroscientific research on the subject started, encouraged by interesting neurological conditions such as blindsight (Weiskrantz 1986) and split brain patients, (Gazzaniga 1988). During this period, theories of consciousness were proposed based on a wide array of research, notably including Baars' 'A Cognitive Theory of Consciousness' in 1988, which is still the basis for much thinking about the neural basis of consciousness. The Association for the Scientific Study of Consciousness was founded as recently as 1994.

Although a growing field, 'consciousness' is not yet an entrenched term in psychology and the cognitive sciences, and many areas of research now co-opted into consciousness research have been pursued successfully using alternative terms. For example, 'unconscious' perception was for a long time, (and still is) referred to as 'subliminal' perception, simply denoting perception that occurs below a threshold (e.g. Holender 1986). 'Unconscious' learning is often described in terms of implicit learning, such that the rules learnt are unreportable by subjects (e.g. Kaufman et al. 2010). Research on attention forms a large part of the current science of consciousness, though it is now argued by some to be a separate phenomenon (see e.g. Koch and Tsuchiya 2007; Lamme 2003, 2004). That is, as suggested many times before, there exist a range of entrenched terms that refer explicitly to the experimental manipulations and tasks used to define a range of different phenomena. Only recently have these phenomena been grouped together, and these terms been traded for the less precise terms of 'conscious' and 'unconscious' processing. Yet aside from the few offered theories of consciousness around (Global Neuronal Workspace Theory,

Integrated Information Theory, etc), consciousness is not an entrenched concept in science, and many still avoid talking about it.

There is however the possibility that it may become entrenched in the future. Several research centers dedicated to the scientific study of consciousness have recently been founded (e.g. the Sackler Center at the University of Sussex), and there are a number of journals dedicated to consciousness (e.g. Consciousness and Cognition). More crucially, the focus on consciousness research from funding bodies and journal editors means that increasing amounts of research is likely to be done, aimed at 'explaining consciousness'.

Yet such an entrenchment of the concept of consciousness would still provide no reason for keeping it. This is because it would be the result of the top-down imposition of the concept from policy-makers, philosophers and phenomenologists, few of whom have an adequate understanding of the methodological problems inherent in this burgeoning field. The focus and hype associated with consciousness research shifts these methodological problems away from the center of discussion, promoting more and more unsolvable debates, and giving rise to yet more deeply flawed inferences being drawn from badly specified experimental work. The concept of consciousness is not essential to cognitive science (there are other, better concepts available), and it would in fact be extremely harmful if it were to become entrenched.

9.3.4.2 The Concept Promotes Stability, Continuity and Generality

If a concept promotes the qualities of stability, continuity and generality in a science, then it might be more disruptive to eliminate it than keep it. Similar to the criterion above, if the elimination of a concept prevents clear communication between scientists, prevents useful comparisons between current and earlier theories from being made, and prevents valuable generalisations from being proposed, then it is usually more productive to keep the concept. However, as seen above, the history of consciousness research has been far from stable. Many taxonomies and theories are relatively recent to the field, and illustrate the widely divergent views on consciousness. Although we have progressively learned more about the mechanisms that underlie the components operationalised in consciousness studies, how consciousness is operationalised and how it is researched periodically switches between two very different methodological approaches (subjective and objective approaches). As research using the concept lacks stability, is not continuous, and lacks consensus on many core issues, eliminating the concept would not radically alter much of the research that is currently done under the banner of 'consciousness science', that can be, and typically is, described in alternative language.

9.3.4.3 The Concept Can Be Used Unambiguously

If a concept does not itself refer to a scientific kind, but to a group of scientific kinds, then its continued use may lead to confusion. For example, if X refers to a group of scientific kinds (a, b, c), then asking questions about X can be interpreted as questions about any of (a, b, c), all of which have different answers. Debates may then ensue in which all sides talk past each other, a particularly unfruitful form of discussion. If X really is ambiguous, and just referring to the scientific kinds a, b and c would avoid problems in communication, then X should be eliminated.

However, it may also be true that in scientific practice these ambiguities are avoided. For questions asked about X in particular fields or contexts, it may be fairly easy to interpret X as referring to one of (a, b, c). If the ambiguities of X are recognised and the concept is used carefully, then it might not be worthwhile to eliminate it. Ereshevsky has used this to argue that although 'species' is ambiguous out of context, its more precise referent (e.g. eco-species) is obvious in the context of particular fields or particular research questions. Therefore, so long as ambiguities are avoided, the concept 'species' does not need to be eliminated. In contrast, Griffiths and Machery have argued that the ambiguity of the concepts 'emotion' and 'concept' is so rife that they are best eliminated. Again, the elimination of a concept is dependent on how it is actually used in scientific practice.

The concept 'consciousness' has often been criticised for being wildly ambiguous (see e.g. Papineau 2003a, b; Sloman 2010; Wilkes 1984, 1988). Consciousness has many different senses in scientific language, so researchers often suggest the particular sense in which they use 'consciousness'. Typically, this is either in terms of phenomenal consciousness vs. reportability of sensory stimuli (e.g. Lamme 2006), consciousness and pre-consciousness (Dehaene et al. 2006) or content and state consciousness (Laureys 2005). While it is hoped that these simple taxonomies of consciousness are sufficient to get everyone talking about the same phenomenon, stark differences in operationalisations of even the same type of consciousness show that even these subtypes are ambiguous (or underspecified) concepts.

As argued above, providing a more detailed taxonomy of consciousness is not likely to help, as it will be a vast and complex one that simply relabels the phenomena identified in the cognitive sciences. In order to spare the debates where researchers simply talk past each other, it was suggested earlier (building on a suggestion from Hulme et al. 2008), that it makes more sense to talk about the specific task used to operationalise consciousness, and to decompose the specific processes and mechanisms used to complete this task. Only by appealing directly to the experimental design and to these components can the ambiguity in claims about 'consciousness' be eradicated. However, this strongly suggests not only that consciousness is inherently ambiguous, but that it is simply unnecessary to describe the phenomena in question.

9.4 Summary

The sections above have argued on a number of fronts that the concept of 'consciousness' should be eliminated from scientific research. One reason for this is the inability to successfully apply scientific methods to concepts of consciousness. Importantly, this means that the heuristic value of such methods for revising theories and generating new research questions is lost as well. Second, scientific kinds can be dissociated, refer to clusters of commonly co-occurring properties that are the product of mechanisms, and support predictions and generalisations. 'Consciousness' and its subtypes do not satisfy these criteria. The severity of these methodological problems also suggest that there is nothing to be gained from treating consciousness (or subtypes of consciousness) as investigative kind concepts. Finally, pragmatic factors that may still count in favour of preserving concepts of consciousness, such as entrenchment, or non-ambiguity, also fail in the case of consciousness. There are methodological, epistemological and practical reasons for eliminating 'consciousness' from cognitive science.

9.5 Why This Account Is Different

At this point it might be questioned how this account differs or goes beyond other radical or eliminativist positions related to consciousness (e.g. Dennett 1991, 1996; Sloman 2007, 2010; Sloman and Chrisley 2003; Wilkes 1984, 1988). While there are clearly similarities between the account offered here and previous work on the scientific explanation of 'consciousness', there are also differences in the motivation and argumentative style, and the empirical support for and implications within this account. These differences ensure that the eliminativist position described in this book offers a new and potentially more solid basis from which to evaluate the science of consciousness.

One way to motivate an eliminativist stance towards a scientific concept of consciousness is to consider the varied ways the concept is used in ordinary language, across cultures and across history. Wilkes uses linguistic considerations as a way of identifying viable targets of scientific research. She argues that 'consciousness' has not had a stable referent over time, does not have a stable referent over languages, and has many different meanings even in western languages. This suggests that it is a recent and culturally relative concept that does not pick out a scientific kind. Despite the commonly held conviction that 'consciousness' refers to an obviously central feature (perhaps *the* central feature) of our mental lives, and picks out a phenomenon that demands (scientific) explanation, the only recent and culture-specific use of this term suggests otherwise.

Illustrative of another approach, Sloman argues that taking a design stance towards 'consciousness' shows that it is a highly ambiguous concept. Instead, questions about 'consciousness' need to be split into more specific questions about the many components that inform our folk concept of 'consciousness'. This includes investigating the architectures necessary for a range of behaviours and abilities and (self) monitoring systems, even modelling how a 'robot philosopher' could come to have our intuitions about consciousness (see Sloman 2007, 2010). By describing the architecture necessary for all the things we associate with consciousness, no further research questions remain.

While these are interesting (and I think viable) positions, the kind of eliminativism followed here is motivated by an assessment of the science of consciousness as it stands, seeing whether it is able to function as a science, and what the status of 'consciousness' is given current empirical evidence. By considering debates about the operationalisation, measurement and taxonomisation of consciousness, and what mechanisms and neural correlates it might be identifiable with, the scientific validity of 'consciousness' can be directly addressed.

This means that the account also comes with empirical support and suggestions for further research 'for free'. By exploring the problems with particular measures, mechanisms and methods, it becomes clear both what *cannot* be claimed, but also what kinds of research questions can be pursued. For example, debates over how to measure consciousness suggest that measures are only valid as measures of particular behavioural responses and of particular sensory and cognitive processes. In arguing that 'consciousness' and its subtypes fail to pick out scientific kinds, the scientific kinds are that are actually relevant to explaining reports and behaviours have been identified. Finally, in critiquing the content-matching strategy, an alternative model based on generative models of perception/cognition provides an explanation of cases where reports and behaviours conflict.

Basing a claim for eliminativism on the current state of consciousness science therefore produces two outcomes. First, a negative claim about the necessity of eliminating 'consciousness' from science can be made, based on the products and needs of actual, current scientific practice. From specific debates found within consciousness science, it has been argued that its methods are inappropriately applied and its research questions confused. Crucially, it has also been argued that preserving the concept of consciousness is detrimental to scientific research. Second, positive claims of what the apparent science of consciousness is really about, as well as what sorts of research questions are valid ones to ask, have been made.

9.6 Eliminating What from Where?

One problem that confronts all similar evaluations of consciousness science is that the resulting positions are just not very satisfactory. Whatever the arguments are in favour of similar positions, the conclusions they draw are just not very palatable. While the claim made here may of course seem radical from the point of view of consciousness science (and those interested in it), the claim is not that consciousness or experience does not exist (whatever that might mean), but that 'consciousness' is not a viable scientific concept. That is, there can be no scientific measures, theories, or mechanisms of consciousness because 'consciousness' is just not a concept that is amenable to scientific methodology.

As discussed in Chap. 8, Dennett's approach is rather different. Instead of eliminating the concept as unusable, he essentially rejects most of the questions and intuitions associated with consciousness, and identifies his Multiple Drafts account of multi-stream information processing as a theory of consciousness. Dennett's account also goes further than Lamme's Neural Stance, which while seemingly revolutionary in how mechanisms and content of consciousness should be identified, in fact assumes a rather standard view of what the contents of consciousness are (and is problematic as a result). Dennett's account accepts the products of cognitive science as providing a theory of consciousness, despite the disparities between this theory and standard assumptions and intuitions made about consciousness.

However, it was argued earlier that it is methodologically clearer to eliminate the concept of consciousness than to promote a theory of it that is radically at odds with the way that many researchers think about consciousness. If the goal of consciousness science is to provide measures and theories that map onto 'consciousness' as we typically think of it, the evidence reviewed in earlier chapters suggests that this goal is not going to be fulfilled. We should instead investigate why we (as western, twentieth century academics) have certain intuitions about 'consciousness', and see these intuitions as the things that require explanation.

For example, in Chap. 7 and 8, concepts related to the contents of visual consciousness were argued to be incompatible with the structure of sensory processing, and a brief account was given of how these concepts might arise given the functions and contents of particular hypothesis generating streams. Our intuitions about the rich contents of visual consciousness reflect the high degree of reliability that our systems of gist processing exhibit in enabling us to move about the world. Models of perception as an inferential process can therefore be used to get at least a rough idea of, for example, why we think we see much more visual detail than we can report, and why we overestimate our abilities to detect changes.

More general concepts such as qualia, or intuitions about the continuous, unified and transparent nature of consciousness, could also be addressed by looking to our cognitive architecture. Dennett addresses some of these intuitions in his Multiple Drafts Theory of consciousness (1991), but Sloman and Chrisley (2003) have provided a model to explain the existence of these philosophical intuitions in a more direct approach. They offer a way of explaining how intuitions about personal, internal, subjective states such as qualia can come about as a product of the different ways we can learn to categorise sensory input, combined with our ability to self-monitor different modes of sensory processing. Lacking a strictly determined set of categories for sensory input, human agents (and other self-organising agents) must generate their own. Although underlying biases mean they will be fairly similar, the categories produced may be different and will reflect the way that input is presented and used by the agent. Different systems of colour words across languages provide a useful example of this (Berlin and Kay 1969). Self-monitoring of internal states is also a feature of reasonably complex agents, through which internal judgements or commentaries on intermediate levels of processing are formed.

Sloman and Chrisley argue that the concept of qualia arises from the conjunction of the two procedures outlined above. That is, qualia talk arises in agents from the use of self-generated concepts to categorise intermediate internal states, as this procedure produces personal, incommunicable, and internally referring concepts. They further conjecture that talk about consciousness is inevitable in agents with this sort of cognitive architecture: "When robots have suitably rich internal information processing architectures some of them will also feel inclined to talk about consciousness, and qualia, in a way similar to the way we do" (Sloman and Chrisley 2003, p. 169). Models such as these can be very useful in explaining how the sorts of philosophical (and now scientific) questions we ask about the mind are predicable given our cognitive structure, yet give rise to misleading scientific research questions. Further work on these kinds of models is necessary to make our mindset about 'consciousness' truly transparent.

Another issue with eliminativist claims is how far their range extends. If consciousness is eliminated from science, it is not clear where this leaves philosophy, and where it leaves folk use of the term. Naturalistic philosophical theories of consciousness that are based on, or derive support from, empirical work, are clearly directly affected by eliminativist claims. If 'consciousness' is not a viable scientific concept, then it cannot figure in empirically informed theories of consciousness. Whether the claims presented here are viewed as relevant to philosophers of mind is a matter for philosophers of mind, but all materialists should presumably feel the pull of arguments that 'consciousness' just is not a phenomenon that has a place in science. Similar to Griffiths' claim that the *scientific* elimination of 'emotion' should also prevent *philosophers* from making theories about 'emotion', it is not clear what the continued philosophical investigation of theories, defining features, functions, and causal powers of consciousness could achieve.

Clearly though, the term consciousness will continue to be used in everyday language, and indeed here it is reasonably clear what is being referred to. Although the term is even more ambiguous than in scientific usage, (including knowing something, explicitly or deliberately doing something, etc), this ambiguity is less problematic than in the sciences. Context typically serves to identify the relevant set of cognitive abilities that are being referred to. For example, whether a lorry driver was conscious while he was driving, even if he cannot remember anything about it, can be easily resolved in folk terms. The lorry driver successfully maneuvered the lorry on the road, so was 'conscious_1', in the sense that he could react appropriately to his environment. However, he is not now 'conscious_2' of his driving, as he cannot remember doing so. Reacting to the environment and remembering doing so are different cognitive abilities implicitly referred to by the folk term of consciousness. They are useful broad categories with which to describe the behaviour of others, and although not particularly well defined (as evidenced in cognitive science), they typically meet the purposes of everyday communication.

Further, experimental philosophy serves to show that folk concepts of consciousness are not the same as philosophical concepts, and have different standards of use. This means that the implications of an eliminativist claim on folk uses of 'consciousness' may be very different from the implications on philosophical uses of the concept. Systma (2010), and Systma and Machery (2010), provide evidence that suggests that 'the folk' do not treat mental states as having phenomenal properties. In this case, the hard problem of consciousness may not be a genuine problem for non-philosophers (despite its overwhelming obviousness to philosophers), and questions about consciousness may well 'shatter' into more specific questions about particular capacities, as suggested above.

In this case, a claim for the elimination of consciousness from science would not necessarily cause any problems for the folk use of the term, as the folk use runs roughly in track with scientific accounts of cognitive/mental phenomena. Interestingly, this may put more pressure on philosophical concepts of consciousness as they can be attacked on two sides; from the cognitive sciences and from experimental philosophy. Further research on how people actually think about consciousness, and how this arises from cognitive as well as from cultural factors, is clearly an important part of any ongoing eliminativist project about consciousness.

References

Baars, B. J. (1988). *A cognitive theory of consciousness*. Cambridge: Cambridge University Press.
Bechtel, W. (2008). *Mental mechanisms: Philosophical perspectives on cognitive neuroscience*. London: Routledge.
Berlin, B., & Kay, P. (1969). *Basic color terms: Their universality and evolution*. Berkeley: University of California Press.
Block, N. (2007). Consciousness, accessibility, and the mesh between psychology and neuroscience. *Behavioral and Brain Sciences, 30,* 481–548.
Boyd, R. (1991). Realism, anti-foundationalism and the enthusiasm for natural kinds. *Philosophical Studies, 61,* 127–148.
Brigandt, I. (2003). Species pluralism does not imply species eliminativism. *Philosophy of Science, 70,* 1305–1316.
Chalmers, D. J. (1995). Facing up to the problem of consciousness. *Journal of Consciousness Studies, 2,* 200–219.
Craver, C. F. (2007). *Explaining the brain: Mechanisms and the mosaic unity of neuroscience*. Oxford: Clarendon.
Dehaene, S., Changeux, J. P., Naccache, L., Sackur, J., & Sergent, C. (2006). Conscious, preconscious, and subliminal processing: A testable taxonomy. *Trends in Cognitive Sciences, 100,* 204–211.
Dennett, D. C. (1991). *Consciousness explained*. London: Penguin.
Dennett, D. C. (1996). Facing backwards on the problem of consciousness. *Journal of Consciousness Studies, 3,* 4–6.
Ereshefsky, M. (1992). Eliminative pluralism. *Philosophy of Science, 59,* 671–690.
Ereshefsky, M. (1998). Species pluralism and anti-realism. *Philosophy of Science, 65,* 103–120.
Ereshefsky, M. (2009). Darwin's solution to the species problem. *Synthese, 175,* 405–425.
Gazzaniga, M. S. (1988). *Mind matters*. Boston: Houghton Mifflin.
Griffiths, P. E. (1997). *What emotions really are: The problem of psychological categories*. Chicago: University of Chicago Press.
Griffiths, P. E. (2004). Emotions as natural kinds and normative kinds. *Philosophy of Science, 71*(5 Supplement: Proceedings of the 2002 Biennial Meeting of the PSA), 901–911.
Holender, D. (1986). Semantic activation without conscious identification in dichotic listening, parafoveal vision, and visual masking: A survey and appraisal. *Behavioral and Brain Sciences, 9,* 1–23.
Hulme, O. J., Friston, K. F., & Zeki, S. (2008). Neural correlates of stimulus reportability. *Journal of Cognitive Neuroscience, 21,* 1602–1610.
Kaufman, S. B., DeYoung, C. G., Gray, J. R., Jiménez, L., Brown, J., & Mackintosh, N. (2010). Implicit learning as an ability. *Cognition, 116,* 321–340.
Koch, C., & Tsuchiya, N. (2007). Attention and consciousness: Two distinct brain processes. *Trends in Cognitive Sciences, 11,* 16–22.

Kornblith, H. (1993). *Inductive inference and its natural ground: An essay in naturalisatic epistemology*. Cambridge: MIT Press.

Lamme, V. A. F. (2003). Why visual attention and awareness are different. *Trends in Cognitive Sciences, 7,* 12–18.

Lamme, V. A. F. (2004). Separate neural definitions of visual consciousness and visual attention; a case for phenomenal awareness. *Neural Networks, 17,* 861–872.

Lamme, V. A. F. (2006). Towards a true neural stance on consciousness. *Trends in Cognitive Sciences, 10,* 494–501.

Laureys, S. (2005). The neural correlates of (un)awareness: Lessons from the vegetative state. *Trends in Cognitive Sciences, 9,* 556–559.

Levine, J. (1983). Materialism and qualia: The explanatory gap. *Pacific Philosophical Quarterly, 64,* 354–361.

Machery, E. (2009). *Doing without concepts*. Oxford: Oxford University Press.

McCauley, R. N., & Bechtel, W. (2001). Explanatory pluralism and heuristic identity theory. *Theory and Psychology, 11,* 736–760.

Milner, A. D., & Goodale, M. A. (2008). Two visual systems re-viewed. *Neuropsychologia, 46,* 774–785.

Papineau, D. (2003a). Could there be a science of consciousness? *Philosophical Issues, 13,* 205–220.

Papineau, D. (2003b). Theories of consciousness. In Q. Smith & A. Jokic (Eds.), *Consciousness: New philosophical essays*. Oxford: Clarendon Press.

Schenk, T., & McIntosh, R. D. (2010). Do we have independent visual streams for perception and action? *Cognitive Neuroscience, 1,* 52–63.

Sloman, A. (2007). Why some machines may need qualia and how they can have them: Including a demanding new Turing test for robot philosophers. In A. Chella & R. Manzotti (Eds.), *AI and consciousness: Theoretical foundations and current approaches AAAI fall symposium 2007.* Technical Report FS-07-01 (pp. 9–16). Menlo Park: AAAI Press.

Sloman, A. (2010). An alternative to working on machine consciousness. *International Journal of Machine Consciousness, 2,* 1–18.

Sloman, A., & Chrisley, R. (2003). Virtual machines and consciousness. *Journal of Consciousness Studies, 10,* 113–172.

Sullivan, J. (2009). The multiplicity of experimental protocols: A challenge to reductionist and non-reductionist models of the unity of neuroscience. *Synthese, 167,* 511–539.

Systma, J. (2010). Folk psychology and phenomenal consciousness. *Philosophy Compass, 5,* 700–711.

Systma, J., & Machery, E. (2010). Two conceptions of subjective experience. *Philosophical Studies, 151,* 299–327.

Ungerleider, L. G., & Mishkin, M. (1982). Two cortical visual systems. In D. J. Ingle, M. A. Goodale, & R. J. W. Mansfield (Eds.), *Analysis of visual behavior* (pp. 549–586). Cambridge: MIT Press.

Weiskrantz, L. (1986). *Blindsight: A case study and implications*. Oxford: Oxford University Press.

Wilkes, K. V. (1984). Is consciousness important? *British Journal for the Philosophy of Science, 35,* 223–243.

Wilkes, K. V. (1988). Yishi, duh, um and consciousness. In A. J. Marcel & E. Bisiach (Eds.), *Consciousness in contemporary science*. Oxford: Oxford University Press.

Wilson, R. A. (1999). Realism, essence, and kind: Resuscitating species essentialism? In R. A. Wilson (Ed.), *Species: New interdisciplinary essays* (pp. 187–207). Cambridge: MIT Press.

Chapter 10
Conclusion

This book has evaluated the scientific viability of the concept of consciousness, and found it to be wanting. By assessing the current theories and methods of consciousness science on their own grounds, and by comparing them to accounts of standard scientific practice from philosophy of science, the argument that 'consciousness' should be eliminated from science goes beyond the argumentative structures typically found in philosophy of mind. Drawing on a broad range of evidence from psychophysics, neurophysiology, psychology and cognitive neuroscience, models of cortical information processing, and philosophy of science, the claims made here are both empirically supported and backed by a rigorous appraisal of scientific methodology. Aside from the purely negative claim found in eliminativist accounts, this empirical grounding also allows positive characterisations to be made about the products of the current science of consciousness, to re-identify real target phenomena and valid research questions for the mind sciences, and to suggest where the intuitions that ground the confused research program on consciousness might come from.

These claims have been reached by investigating the ways in which a number of methods fail to be adequately applied in consciousness science, both in particular cases and in fulfilling their standard heuristic role. Problems associated with the use of both subjective and objective measures have no methodologically valid solutions. Proponents of phenomenological and introspective techniques to gain 'accurate' subjective reports fail to address the issue of response bias, as formalised within Signal Detection Theory. Second order confidence ratings may simply measure meta-cognitive capacity. While objective measures such as sensitivity d' are devoid of response bias, they may simply measure information processing. Qualitative difference and dissociation methods cannot resolve which type of measure to use. Dissociation methods must be used within an interpretive framework that refers to specific experimental manipulations, suggests new research questions, and that is itself revised in light of further dissociation paradigms. Concepts of consciousness prevent the valid application of dissociation methodology, and its essential heuristic value is lost.

Later sections showed how integrative methods fail to identify clusters of behavioural and neurophysiological measures of consciousness, but instead identify

E. Irvine, *Consciousness as a Scientific Concept,* Studies in Brain and Mind 5,
DOI 10.1007/978-94-007-5173-6_10,

clusters of measures of sensory and cognitive processes. The way that demarcation criteria are inadequately applied to neural mechanisms of consciousness show again that mechanisms, and components of mechanisms, can only be identified for sensory and cognitive processes. Both of these approaches show that 'consciousness' does not refer a scientific kind, or a group of related scientific kinds, so cannot even function as a useful umbrella term. There is little in common between the wide range of ways that consciousness is operationalised at the behavioural, mechanistic, or functional level.

Finally, it was argued that the content-matching approach shows how the properties attributed to conscious perceptual content are in fact deeply incompatible with the structure of perceptual processing. Sensory memory does not provide a set of rich phenomenal content, and multi-stream inferential processing is incompatible with many accounts of conscious content. Taking seriously the attempts to correlate (and identify) the contents of consciousness with the contents of neural processes leads, at best, to a drastic revision of the concept of 'the contents of consciousness', and one in which core features of the original target phenomenon disappear.

These chapters all reach similar conclusions; it is not possible to successfully apply scientific methods to concepts of consciousness, and successful applications of these methods show that there are no phenomena, no mechanisms, no natural kinds, and no content, that could be referred to by 'consciousness' that are not already more clearly described in the cognitive sciences. Consciousness science obscures this fact by using vague terms to group phenomena that are recognised elsewhere as being quite different.

By looking at methodological, epistemological and pragmatic criteria used to state scientific eliminativist claims, it becomes increasingly clear that not only does 'consciousness' fail to be a useful scientific term, but it is also a harmful one to scientific practice. The use of the concept 'consciousness' prevents standard scientific methods form being adequately applied, inhibits standard research heuristics, does not allow generalisations and predictions to be made, prevents clear communication across research groups, and thus forces research into unproductive and confused directions. Instead, research should be directed back at the well-framed research questions of the cognitive sciences, and to the question of why (perhaps only a limited number of researchers) have such strong intuitions about this confusing concept.

As consciousness is both a trendy and a touchy topic, eliminativist claims are either viewed with incredulity by those working in the field, or given a nod of acceptance by those who steer clear of it. There are rarely converts. However, if nothing else the arguments laid out in this book are intended to provide some *scientific* reasons, rather than linguistic, historical, conceptual or intuitive reasons, for the elimination of concepts of consciousness from science, and from related naturalistic thinking. Even if these are not accepted as a definitive reasons to eliminate 'consciousness' from scientific discourse, they should at least illustrate the methodological problems that consciousness science currently faces, and highlight that they are more serious than currently acknowledged.

Finally, while eliminativism is often seen as a purely negative stance, scientific eliminativism is both necessary to the progress of science, and must also offer a positive contribution in terms of providing conceptual clarification, explanations of current results, and future research questions. Far from offering a derogatory characterisation of the claims of current science and philosophy centered on consciousness, this book is based on taking these claims seriously, seeing if they are methodological justified, and offering alternative accounts. Unfortunately (for some), a science of consciousness is not possible. On the other hand, it turns out that one is not necessary to answer the real questions we can pose about our mental life. This approach from philosophy of science appears to be a powerful one, and importantly a productive one, and will, I hope, provoke a new wave of discussions about what consciousness science can actually achieve.

Appendix: Dice Game

The task involves three dice, two of which are normal and carry the numbers 1–6, one number on each face, while the third dice has three faces covered in zeros, and the other three faces covered in threes. An experimenter rolls the three dice out of sight of the subject and reports the sum of the dice to the subject. Given only the knowledge about the format of the dice and the sum of the three dice on this occasion, the subject then has to say whether the 'abnormal' die shows a zero or a three.

The subject can do this (either explicitly or implicitly) by figuring out the conditional probabilities of all the possible sums, given that the dice either shows a zero or a three. For example, the probably of the sum being 2 *if the 'abnormal' dice shows a three* is 0 (the only way of getting a sum of 2 is to have a one-one-zero combination). The probability of the sum being 2 *if the 'abnormal' dice shows a zero* is 1/36 (there is only one way this can happen, and there are 36 distinct combinations using a zero). The best response given a sum of 2 is therefore to say that the 'abnormal' dice shows a zero. By comparing the probabilities of a sum conditional on the 'abnormal' dice showing either a zero or a three, the optimal response can be ascertained for all sums. The conditional probabilities for all the sums are given in the left-hand columns in the table below.

As you can see from Table A.1, most of the probability distribution for the 'zero' column is clustered around the smaller sums, while most of the probability distribution for the 'three' column is clustered around larger sums. This can be used to establish the optimal response criterion to use in order to maximise the number of correct responses in the Type 1 task. In this example, the optimal response strategy is to say that for all sums below and including 8, the 'abnormal' dice shows a zero, and for all sums over and including 9, that the dice shows a three. That is, the most likely way that a sum up to and including 8 can occur is by the abnormal dice showing a zero, and for any higher sums that it shows a three. This response criterion (respond 'three' above sums of 8) uses all of the available information and is the optimal response criterion to use.

In order to compute the probabilities relevant for a Type 2 response, the subject must figure out how likely the sum was, this time conditional on whether their response (based on their response criterion) was correct or incorrect. If a sum is more likely to have occurred if their Type 1 response was correct, then the subject will respond that they are confident in their Type 1 response. Conversely, if a sum is

E. Irvine, *Consciousness as a Scientific Concept,* Studies in Brain and Mind 5,
DOI 10.1007/978-94-007-5173-6, © Springer Science+Business Media Dordrecht 2013

Table A.1 Probability distribution for Type 1 and Type 2 responses. (Adapted from Table 2.1, Galvin et al. 2003, p. 850, and reprinted with permission from Springer Science and Business Media)

Sum	Type 1 Probability of sum given a zero on the 'abnormal' dice	Type 1 Probability of sum given a three on the 'abnormal' dice	Type 2 Probability of sum given correct answer	Greater/lesser	Type 2 Probability of sum given incorrect answer
2	1/36	0	1/52	>	0
3	2/36	0	2/52	>	0
4	3/36	0	3/52	>	0
5	4/36	1/36	4/52	>	1/20
6	5/36	2/36	5/52	<	2/20
7	6/36	3/36	6/52	<	3/20
8	5/36	4/36	5/52	<	4/20
9	4/36	5/36	5/52	<	4/20
10	3/36	6/36	6/52	<	3/20
11	2/36	5/36	5/52	<	2/20
12	1/36	4/36	4/52	>	1/20
13	0	3/36	3/52	>	0
14	0	2/36	2/52	>	0
15	0	1/36	1/52	>	0
Total	1	1	1		1

more likely to have occurred if their Type 1 response was incorrect, then the subject will respond that they are *not* confident in their Type 1 response. Good performance at the Type 2 task consists in being able to judge whether a Type 1 response was correct or not. As explained above, this means that subjects who are aware of the stimuli (in this case the sum), should be highly confident in their correct responses, and have low confidence in their incorrect responses.

For example, given the response criterion described above, the subject will always say that the 'abnormal' dice shows a three for the sum of 13. The probability of the sum being 13 conditional on the subject giving the correct Type 1 response is more likely than its converse, as the only way to get a sum of 13 is to have a three on the 'abnormal' dice, and this is what the subject will always say, given their response criterion. Given that the subjects' responses in the Type 1 task in this example are based on an optimal response criterion, it seems that subjects would always be able to tell when they had given correct and incorrect responses, thereby giving them high levels of performance on the Type 2 task. However, this is not the case.

The reliability of Type 2 judgements for sums of 13 and above, or 4 and below, are high because the subject will always be correct in their Type 1 response, given their response criterion. However, for all the other sums, the Type 2 judgements are more complicated as subjects can be wrong in their Type 1 response. For example, for the sum of 10, the subject will again use their optimal response criterion to respond that the 'abnormal' dice shows a three. However, the sum of 10 can be reached both by having the 'abnormal' dice show a three or a zero, so the subject can be wrong in their Type 1 response. In fact, the probability of the sum being 10 conditional on the subject being correct (i.e. that the 'abnormal' dice shows a three) is *less* than the probability of the sum conditional on the subject being incorrect. The probabilities of the sum being 10 conditional on a subject being correct or incorrect are shown in the right hand table above (see Galvin et al. 2003 for more details).

From the table, it can be seen that the probabilities of a sum conditional on the subject giving the correct answer are *lower* than the probabilities of the sum conditional on the subject giving the incorrect answer for the sums 6–11. Given that these sums are the most common ones (occurring 69 %) of the time, this means that although the subject gives the correct answer most of the time in the Type 1 task, he has a significantly compromised ability to accurately judge the correctness his Type 1 responses—the essence of the Type 2 task. Using all the stimulus information to hand to generate the Type 2 conditional probabilities, the subject should say he is wrong in his Type 1 responses for the sums 6–11 even though he will actually be right most of the time. This peculiar result reflects the features of the probability distribution of the task. Although the subject is able to process and judge information optimally, just the statistical features of the example ensure that Type 2 performance will be worse than Type 1 performance.

Reference

Galvin, S. J., Podd, J. V., Drga, V., & Whitmore, J. (2003). Type 2 tasks in the theory of signal detectability: Discrimination between correct and incorrect decisions. *Psychonomic Bulletin and Review, 10,* 843–876.

Index

E. Irvine, *Consciousness as a Scientific Concept*, Studies in Brain and Mind 5,
DOI 10.1007/978-94-007-5173-6,

MIX
Papier aus verantwortungsvollen Quellen
Paper from responsible sources
FSC® C105338

If you have any concerns about our products,
you can contact us on
ProductSafety@springernature.com

In case Publisher is established outside the EU,
the EU authorized representative is:
Springer Nature Customer Service Center GmbH
Europaplatz 3, 69115 Heidelberg, Germany

Printed by Libri Plureos GmbH
in Hamburg, Germany